Upstream Engineering and Flow Assurance Forum 2014

Core Programming Area at the 2014 AIChE Spring Meeting and 10th Global Congress on Process Safety

New Orleans, Louisiana, USA
30 March – 3 April 2014

ISBN: 978-1-63439-081-1

Printed from e-media with permission by:

Curran Associates, Inc.
57 Morehouse Lane
Red Hook, NY 12571

Some format issues inherent in the e-media version may also appear in this print version.

Printed by Curran Associates, Inc. (2014)

For permission requests, please contact AIChE
at the address below.

AIChE
3 Park Avenue
New York, NY 10016-5991

Phone: (203) 702-7660
Fax: (203) 775-5177

www.aiche.org

Additional copies of this publication are available from:

Curran Associates, Inc.
57 Morehouse Lane
Red Hook, NY 12571 USA
Phone: 845-758-0400
Fax: 845-758-2634
Email: curran@proceedings.com
Web: www.proceedings.com

TABLE OF CONTENTS

354012 Possible vs. Practical: Engineers Must Lead the Development of Practical Technologies

Abstract:

Chemical Engineering is need today more than ever. There is a strong societal desire for sources of energy and feedstocks that can sustainably support a high standard of living for all. Thermodynamic, material and economic challenges stand in the way. Most of the general public does not have the basic understanding of the energy flows, chemical transformations, and the scale of the systems that support our current way of life. These concepts are fundamental to chemical engineering. Chemical engineers should be on the vanguard of separating hype from truly practical options yet occationally we let society down. This talk will present several examples where the fundamentals of chemical engineering can be used to access the practicality of proposed solutions to energy and water issues, as well as explain why the venture capital model struggles in the materials market.

352091 A Model for Three Phase Flow in Horizontal Pipes

Abstract:

Experimental observations on three phase flow of gas and two immiscible liquids in pipes is very limited inspite of its prevalence in the oil and gas industry. A few investigators reported observations on the regime maps, phase hold-ups and pressure drop. Extension of the two phase flow models to explain three phase flows by defining two out of the three phases as a pseudo-homogeneous phase was inadequate. In this paper, an equation for the pressure drop in three phase flow through horizontal pipes is derived by extending the separated flow model for the three phases in terms of Lockhart-Martinelli parameters. The model is validated with the experimental data available in the literature.

353112 Immiscible Displacement in Homogeneous and Heterogeneous Porous Media: Quantifying Interfacial Phenomena

Abstract:

Immiscible displacement is ubiquitous in many subsurface multiphase flow systems and pore-scale characterization of these processes aids in the development of quantitative, empirical relationships required in Darcy-scale models. Here, we present experiments in which a wetting fluid is displaced by a non-wetting fluid in homogeneous and heterogeneous porous micromodels. The resulting steady-state fluid distributions are analyzed to characterize fluid-fluid and fluid-solid interfacial length in terms of saturation, fingering pattern, capillary number, and porosity. A comparison between the homogeneous and heterogeneous experiments demonstrates that heterogeneity expands the range of conditions under which capillary fingering is observed. In all micromodels, the number of non-wetting isolated fingers (i.e., blobs and ganglia) increased within the range of injection flow rates. The results show a clear relationship between interfacial length and saturation; however we did not observe a clear relationship between interfacial length and fingering pattern. In addition, once saturation was taken into account, interfacial length was independent of the capillary number. A clear relationship is observed between the non-wetting–wetting fluid interfacial length, non-wetting fluid saturation, and porosity. We anticipate our results will serve as model to quantify relationships for interfacial phenomena in pore-scale drainage and help with the development of Darcy-scale models.

347844 Comparison Between Two-Phase Flow Models and Wire Mesh Sensor Measurements in Medium and Large Diameter Pipes

Abstract:

Wire mesh sensors (WMS) are recently developed fast imaging instruments that have been used to investigate two-phase flows. WMS measure either the electrical conductivity or capacitance of the two phases, allowing the local void fraction to be inferred [1, 2]. The WMS have been benchmarked against the gamma densitometer, a high-resolution instrument widely used in the industry for two-phase flow metering; therefore providing confidence with experimental data [3]. In this work, we compare the results of an extensive WMS experimental campaign,with a number of two-phase flow models based on the drift-flux approach [4]. Void fraction is a critical parameter in two-phase flows. Drift-flux models are semi-mechanistic models that compare to fully mechanistic models use an explicit equation for the void fraction, thus reducing computational time significantly. Moreover, most mechanistic models cannot capture all the flow patterns and pipe inclinations within a single framework without adding complexity and computational overhead. A number of authors have developed drift-flux models for a range of flow patterns and inclinations which seem suitable for bubbly, slug and churn flows [5, 6].

The experiments consisted of WMS measurements using water/air and oil/air in medium and large diameter pipes with pipe inclinations ranging from vertical to horizontal. We compare the performance of several drift-flux closure equations developed for a range of flow patterns and inclinations. The models have been ranked and the empirical parameters of the best performers have been optimized. Additional comparisons between the final optimized models, the experimental data and the mechanistic model OLGA-S are also presented in this study.

[1] Prasser et al. (1998) A new electrode-mesh tomograph for gas-liquid flows, *Flow Meas. Instrum.*,9, 111-119

[2] Da Silva et al. (2007) Capacitance wire-mesh sensor for fast measurement of phase fraction distribution, *Meas. Sci. Technol.*, 18, 2245-2251

[3] Sharaf et al. (2011) Comparison between wire mesh sensor and gamma densitometry void measurements in two-phase flow, *Meas. Sci. Technol.*, 22, 1-13

[4] Ishii and Hibiki (2006) Thermo-fluid Dynamics of Two-Phase Flow, Springer

[5] Shi et al. (2005) Drift-flux Modeling of Two-Phase Flow in Wellbores, *SPE J.*, 24-33

[6] Choi et al. (2012) An Efficient Drift-flux Closure Relationship to Estimate Liquid Holdups of Gas-Liquid Two-Phase Flow in Pipes, *Energies*, 5, 5294-5306

351674 Experimental Flow Loop Setup Used to Investigate Hydrate Formation in Water-in-Oil Emulsions

Abstract:

Gas hydrates are non-stoichiometric crystalline solids with a characteristic guest – host structure that develop from mixtures of water and low molecular weight hydrocarbons. The agglomeration and plug formation tendencies of hydrates in oil and gas flow lines and the ensuing capital losses have provided an impetus to develop reliable, efficient, and economical flow assurance strategies. In this work, we present an experimental setup that is used to quantify emulsion behavior and hydrate formation in flowing systems. The flow loop (8 ft X 8 ft) consists of jacketed 1 in. tubing. Fluid is circulated in the flow loop using a 5 hp progressive cavity pump. The process fluid temperature is controlled in the range (-30°C to 150°C). Flow measurement is accomplished using a mass flow meter, and an inline microscope provides quantitative drop size distributions of emulsions as well as hydrate particle size. The microscope also provides real time images of hydrate slurry flow, adhesion of hydrate particles to inner pipe walls, and eventual plug formation. Quantitative droplet/particle size data are utilized to establish relationships between emulsion droplet size and hydrate behavior in flowing conditions for flow assurance applications.

astro technology

ADVANCED DEEPWATER MONITORING

Reza Asgharzadeh Shishavan

John Hedengren

LAND | DEEPWATER | AEROSPACE | astrotechnology.com

astro technology

ABOUT ASTRO TECHNOLOGY

ADVANCED INSTRUMENTATION FOR:

- Subsea fields
- Pipelines and risers
- Space structures
- Rocket Motors

ENGINEERING CAPABILITIES INCLUDE:

- System integration
- Real-time embedded systems
- Experimental stress analysis
- Fiber-optic sensor technology
- Conventional sensor integration
- Environmentally hardened systems
- Software development

LAND | DEEPWATER | AEROSPACE | astrotechnology.com

astro technology

FAILURES TO MONITOR AND PREDICT

- Detect early warning signs
- Automate monitoring of critical systems
- Give critical data to key decision makers

Deepwater Horizon
2010

Thunder Horse
2005

Texas City Refinery
2005

LAND | DEEPWATER | AEROSPACE | **astrotechnology.com**

astro technology

DEVIL'S TOWER BASS LITE

- Located in Atwater Valley Block 426
- Utilizes 20.3-cm (8-in) diameter flowline
- 90-km (56-mi) length
- Ties to Devil's Tower in Mississippi Canyon
- Production – Up to 130 million cubic feet per day
- Water depth – 2,050 m (6,750 feet)
- Commenced operation in February 2008

MONITORING:

- Pressure
- Temperature
- Hoop and Axial Strain

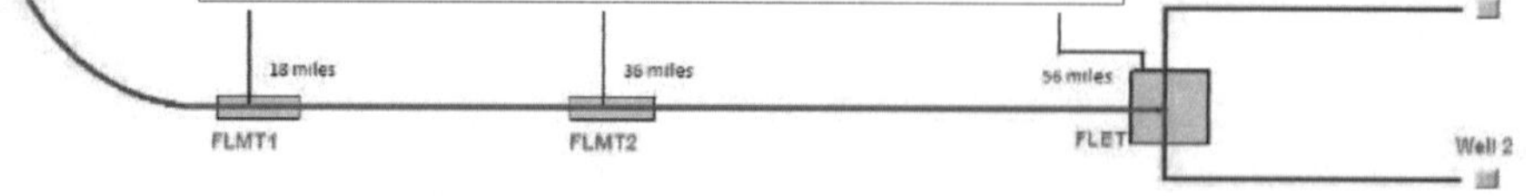

LAND | DEEPWATER | AEROSPACE | **astrotechnology.com**

astro technology

PREVIOUS INSTRUMENTATION ON DEEPWATER RISERS

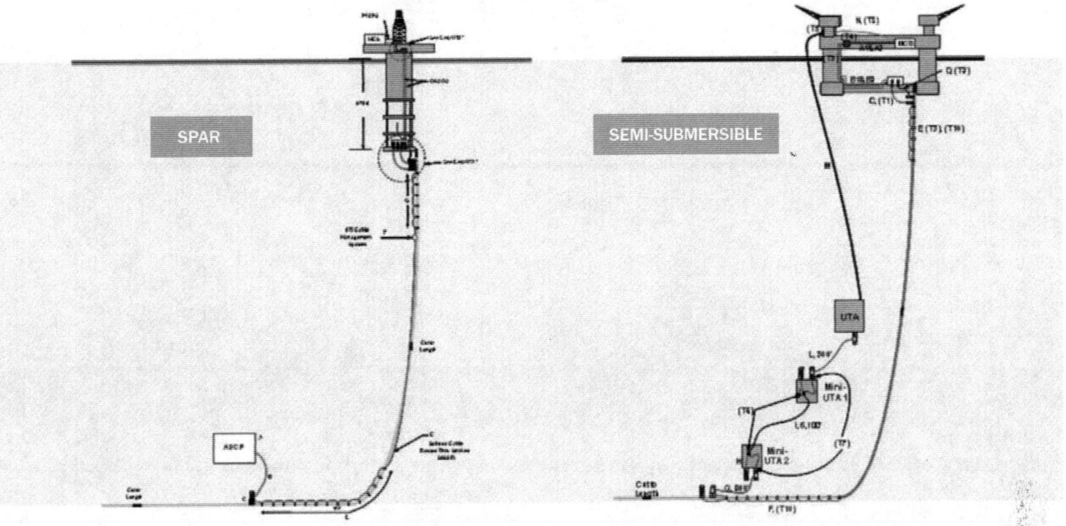

LAND | DEEPWATER | AEROSPACE | **astrotechnology.com**

astro technology

TROIKA — GULF OF MEXICO

ASTRO TECHNOLOGY pioneered the use of fiber-optic sensors on a subsea pipeline to monitor pressure, strain and vibration in external casing pipe bundle during fabrication.

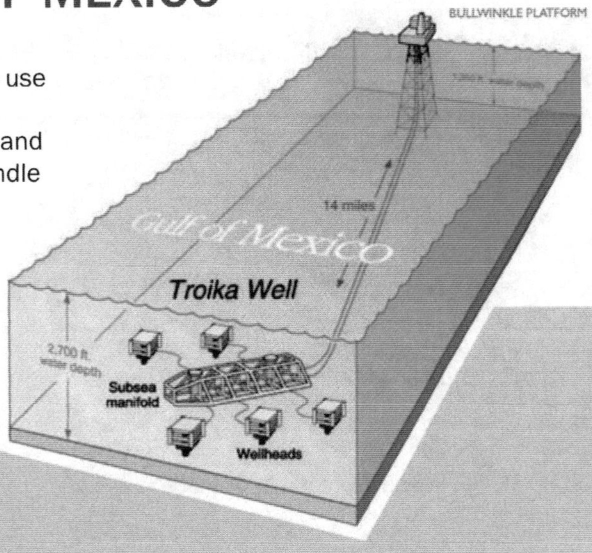

LAND | DEEPWATER | AEROSPACE | **astrotechnology.com**

astro technology

FIBER-OPTIC SENSORS FOR DEEPWATER DRILLING
OCEAN CLIPPER

LAND | DEEPWATER | AEROSPACE | **astrotechnology.com**

astro technology

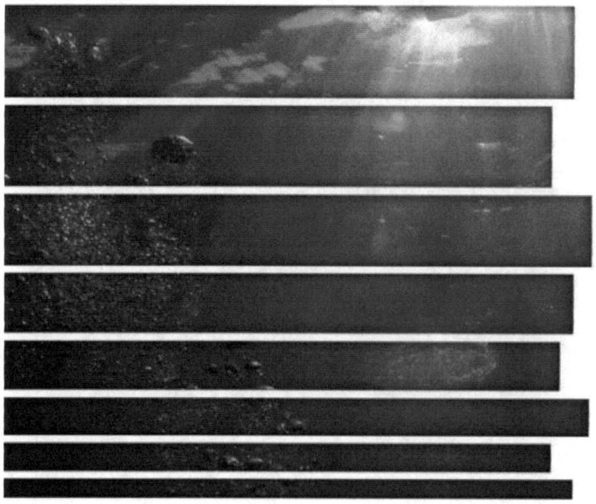

OVERVIEW *of* INSTRUMENTATION METHODS

TRIDENT
SUBSEA SYSTEMS
ASTRO TECHNOLOGY
DEEPER INSIGHT.

LAND | DEEPWATER | AEROSPACE | **astrotechnology.com**

astro technology

FIBER BRAGG GRATINGS

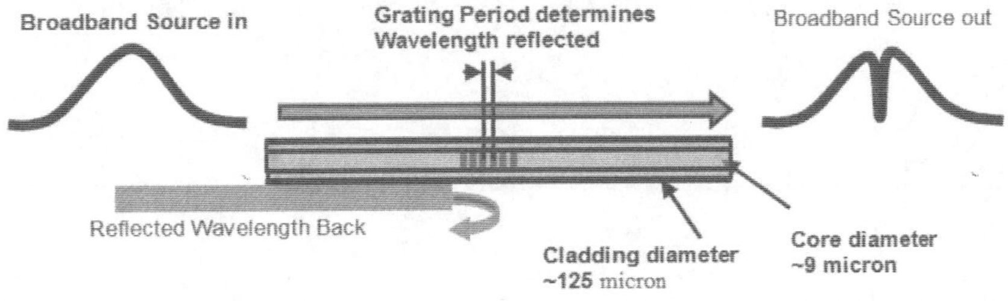

Relationship between Strain (ε) and Change in Wavelength ($\Delta\lambda_b$)

$\Delta\lambda_b / \lambda_b = (1-p_e)\varepsilon$, where p_e is the photoelastic constant for glass and λ_b is the base wavelength

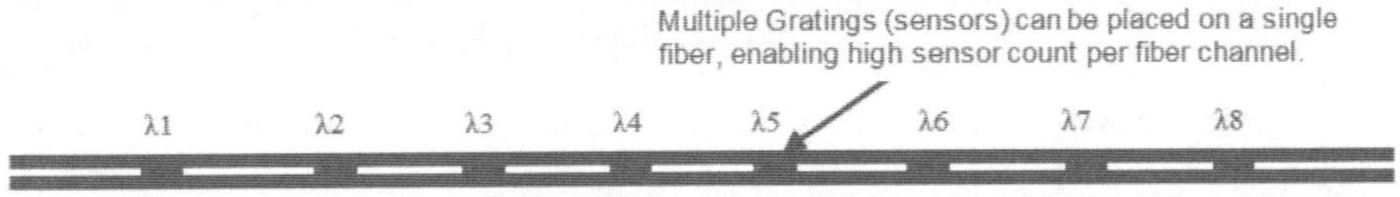

LAND | DEEPWATER | AEROSPACE | **astrotechnology.com**

astro technology

RELATIONSHIP BETWEEN WAVELENGTH & STRAIN

EXAMPLE BASE WAVELENGTH FOR A SINGLE FBG

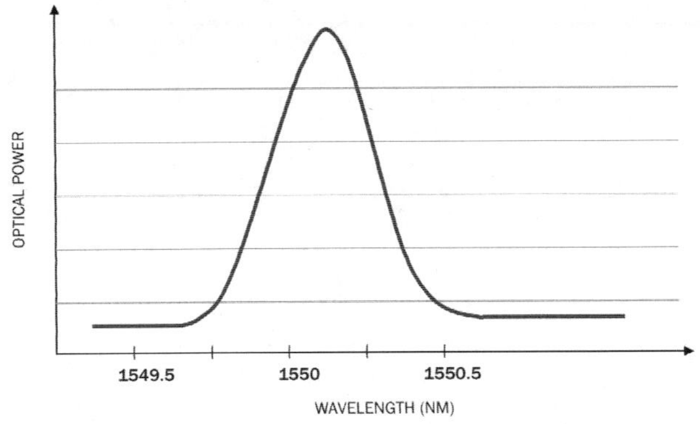

LAND | DEEPWATER | AEROSPACE | **astrotechnology.com**

astro technology

RELATIONSHIP BETWEEN WAVELENGTH & STRAIN — TENSION

REFLECTED SIGNAL FROM THE PREVIOUS FBG IN TENSION

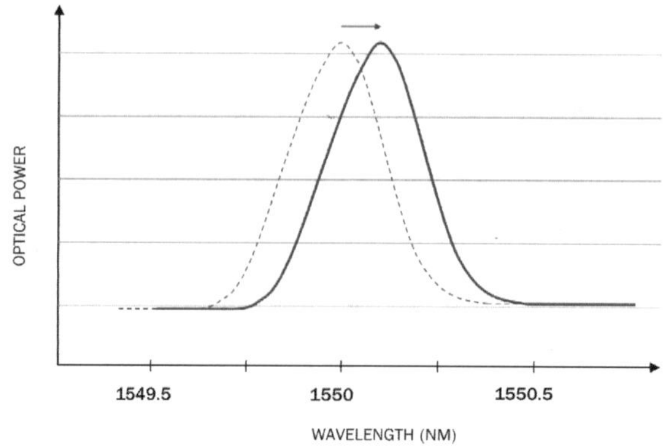

TRIDENT

LAND | DEEPWATER | AEROSPACE | astrotechnology.com

astro technology

RELATIONSHIP BETWEEN WAVELENGTH & STRAIN — COMPRESSION

REFLECTED SIGNAL FROM THE PREVIOUS FBG IN COMPRESSION

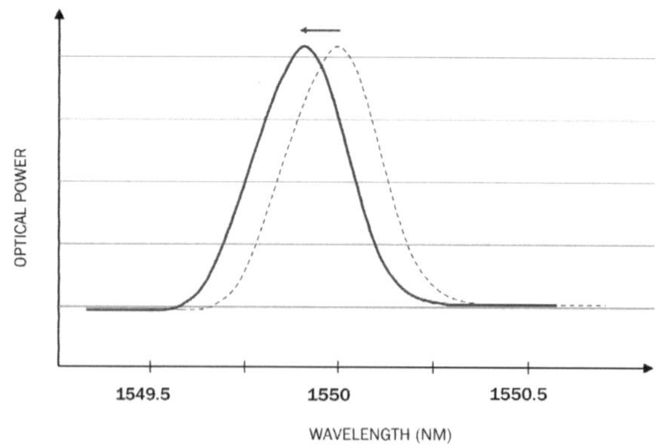

TRIDENT

LAND | DEEPWATER | AEROSPACE | astrotechnology.com

TENSION LEG PLATFORM SENSORS

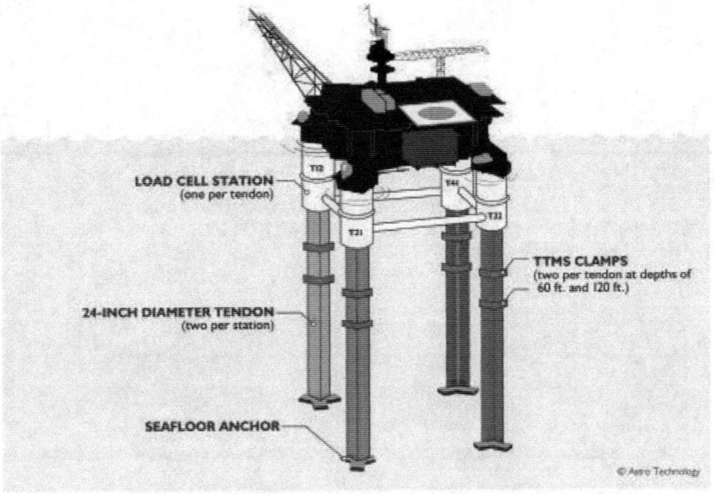

LAND | DEEPWATER | AEROSPACE | **astrotechnology.com**

TENDON TENSION MONITORING SYSTEM

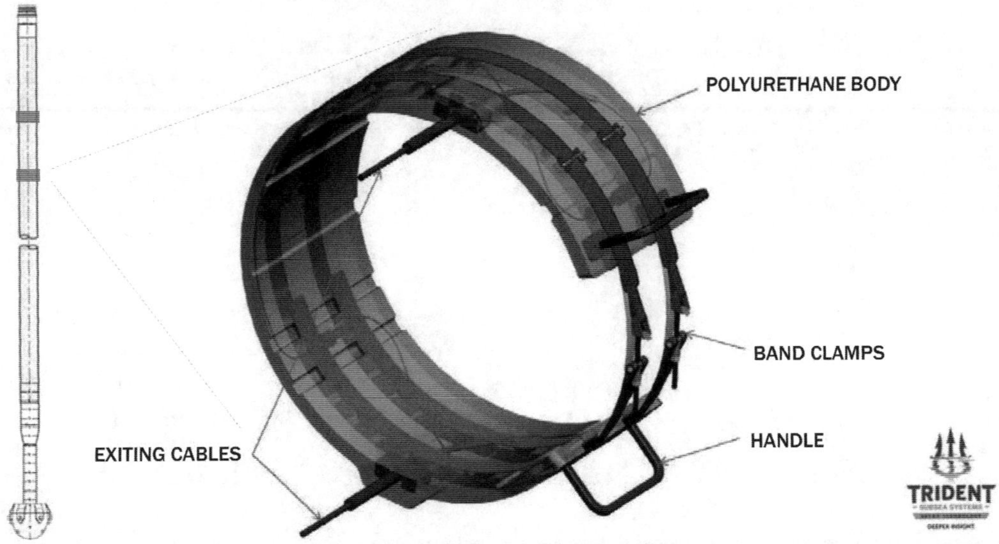

LAND | DEEPWATER | AEROSPACE | **astrotechnology.com**

astro technology

TENDON BAND PREPARATION

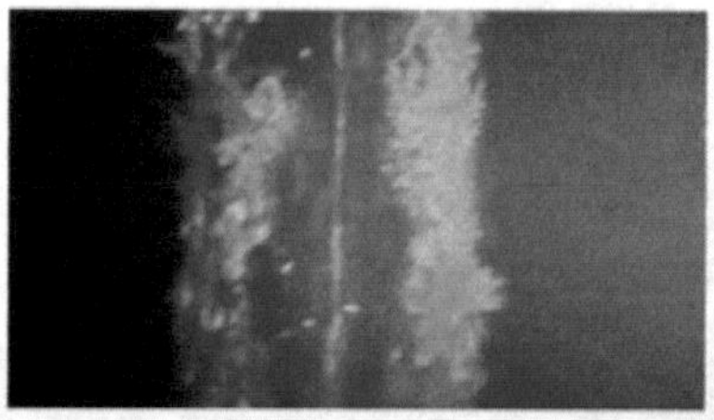

Marine Growth
(BEFORE)

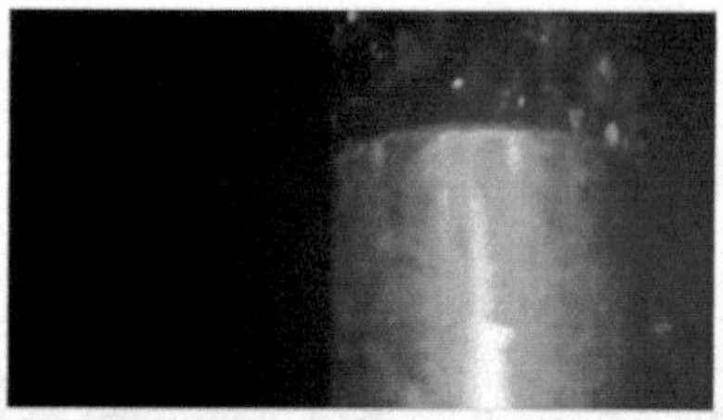

Clean Band
(AFTER)

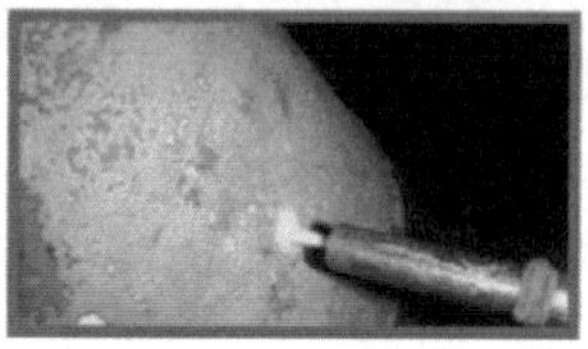

Cleaning with Water Jet

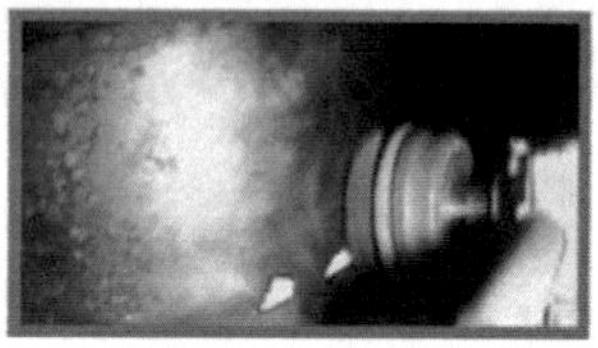

Polishing to Bare Metal

TRIDENT

LAND | DEEPWATER | AEROSPACE | astrotechnology.com

astro technology

DIVER INSTALLATION

Diver with Clamp

Riser Preparation

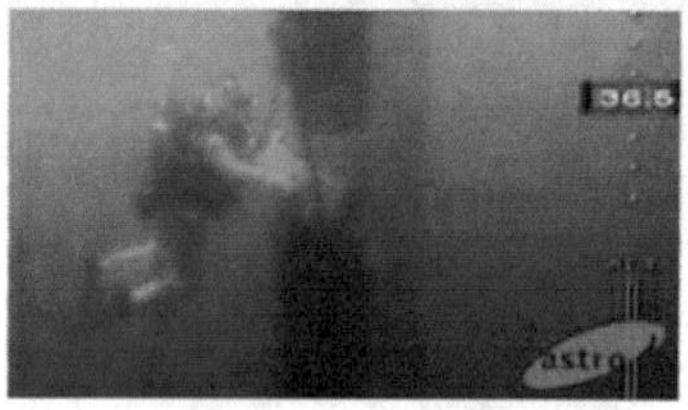

Clamp Installation

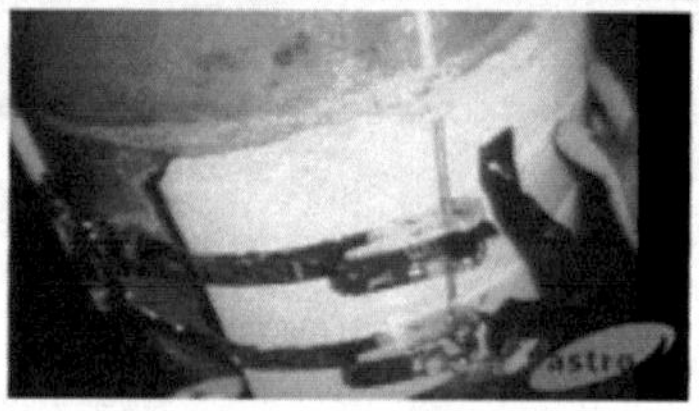

Clamp Inspection

TRIDENT

LAND | DEEPWATER | AEROSPACE | astrotechnology.com

TEMPERATURE & STRAIN GAUGES

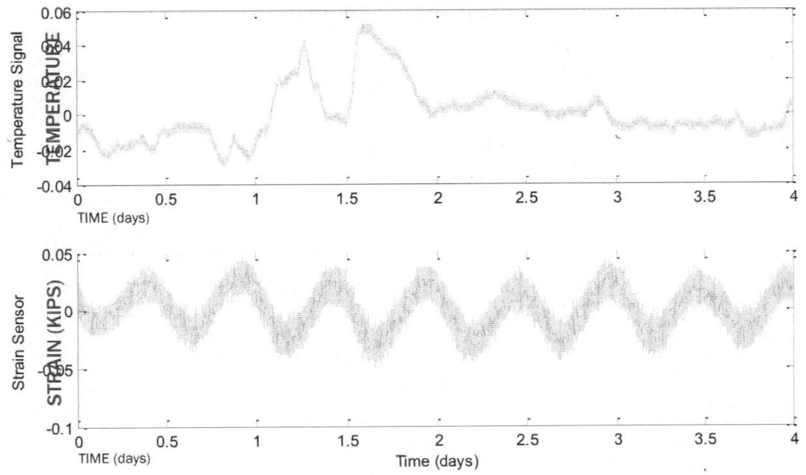

LAND | DEEPWATER | AEROSPACE | **astrotechnology.com**

COMPENSATE FOR DYNAMIC TEMPERATURE VARIATION

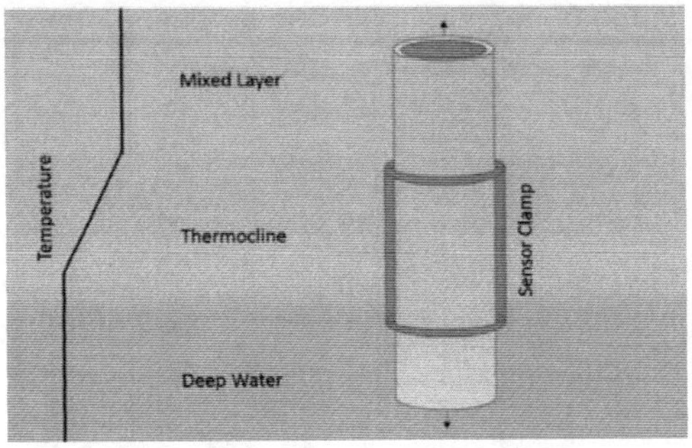

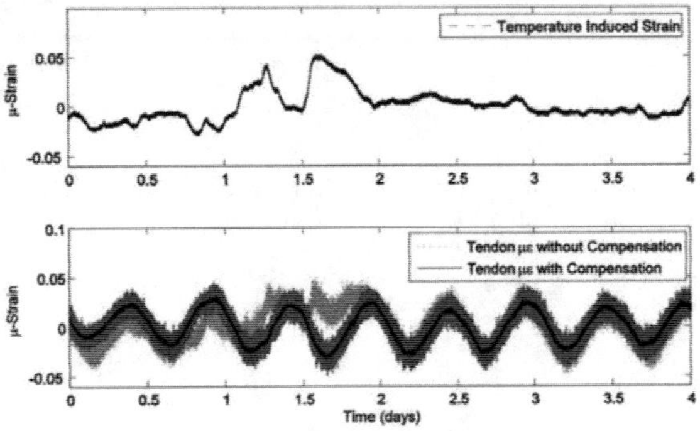

LAND | DEEPWATER | AEROSPACE | **astrotechnology.com**

astro technology

TRIDENT MONITORING SYSTEM

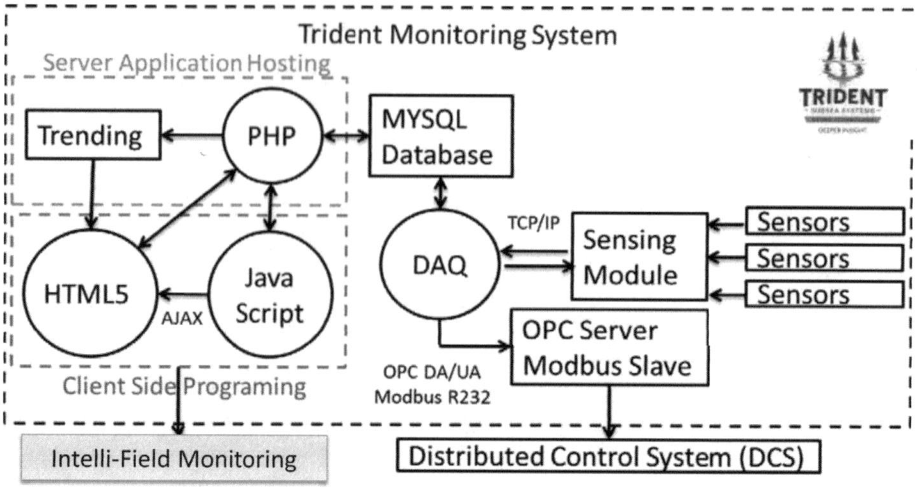

astro technology

ENSURE SUBSEA BONDING STRENTH

Button Pull Tests

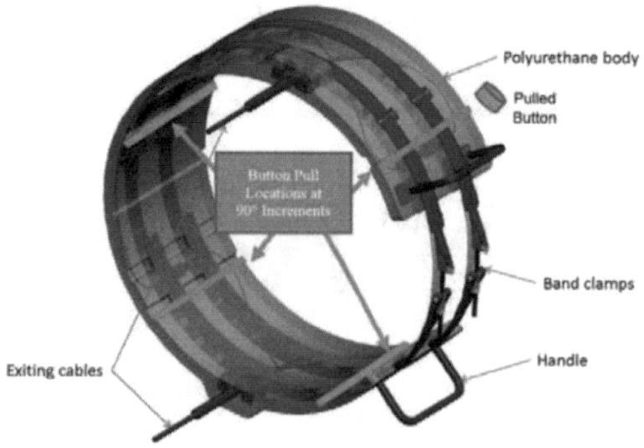

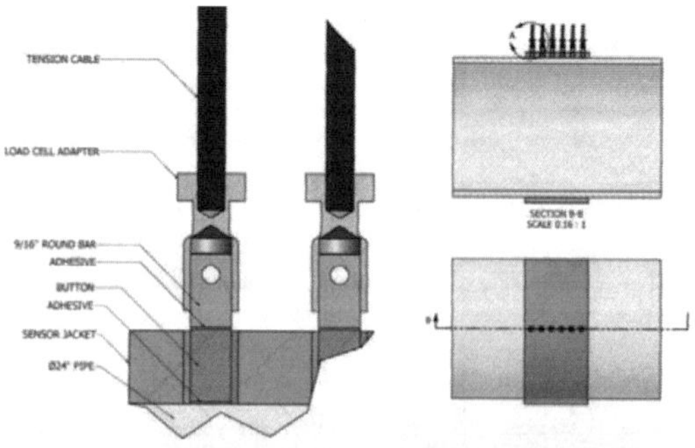

BONDING MAINTAINED IN ACCELERATED AGING

Clamp	Avg (psi)	StDev (psi)	Min (psi)	Max (psi)	Description
BB	**292.0**	108.9	113.6	498.6	**Dry Bonded Control**
AA	**81.4**	47.2	31.1	193.5	**Wet Bonded Control**
A	94.0	30.4	45.0	155.8	Four Point Bending (Left)
B	174.1	112.4	29.8	503.3	Four Point Bending(Left)
E	**91.1**	46.9	49.5	150.8	**Four Point Bending (Center)**
C	142.5	65.6	45.8	267.9	Four Point Bending (Right)
D	136.3	73.6	57.5	358.7	Four Point Bending (Right)
F	**114.0**	53.7	33.6	225.7	**Tension**
G	**105.3**	59.8	29.7	268.8	**Compression**

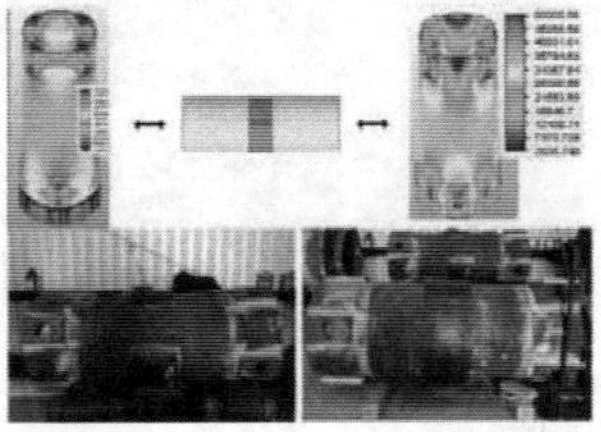

IMPROVED CLAMP DESIGN

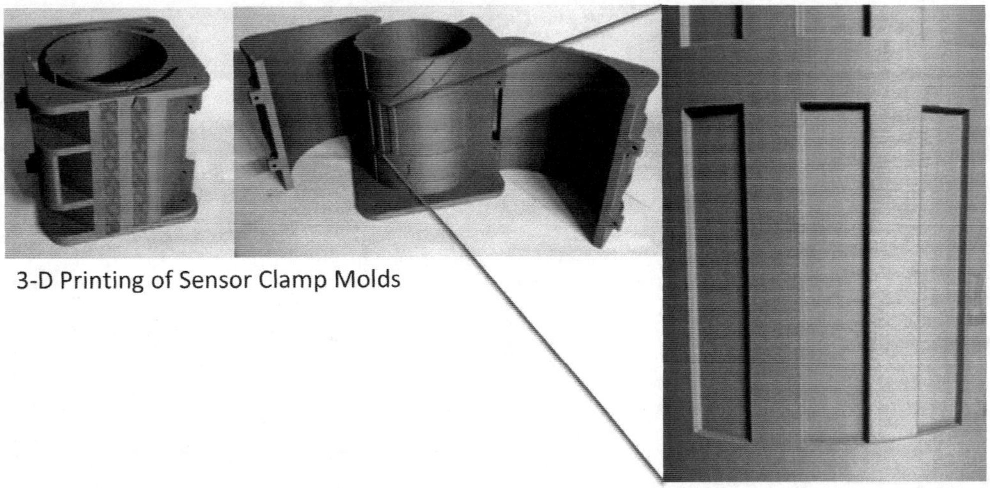

3-D Printing of Sensor Clamp Molds

LAND | DEEPWATER | AEROSPACE | astrotechnology.com

astro technology

BOAT DOCKING

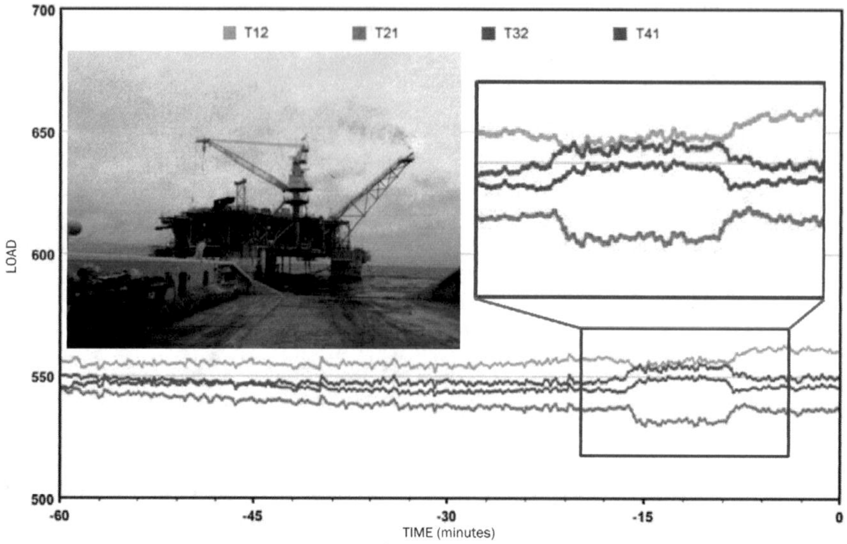

LAND | DEEPWATER | AEROSPACE | astrotechnology.com

astro technology

WEST AFRICAN ENVIRONMENT:

SUDDEN AND POTENTIALLY VIOLENT SQUALLS

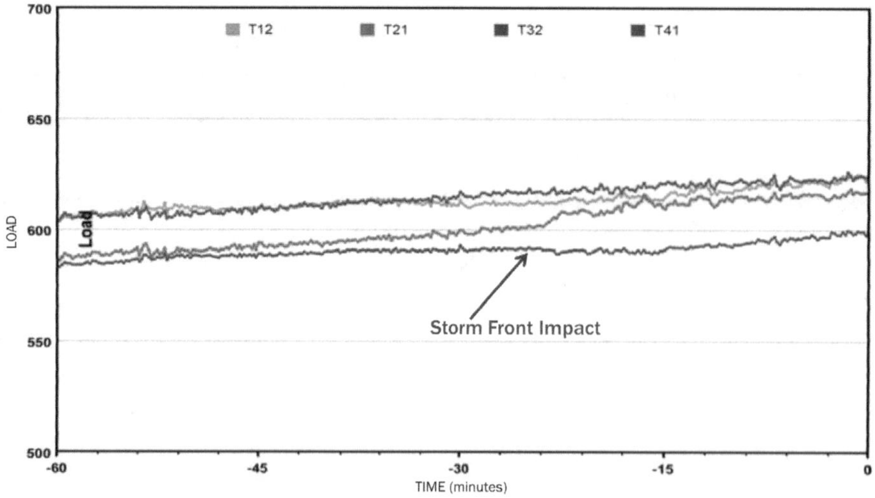

LAND | DEEPWATER | AEROSPACE | astrotechnology.com

SENSOR CALIBRATION WITH TIDE CYCLES

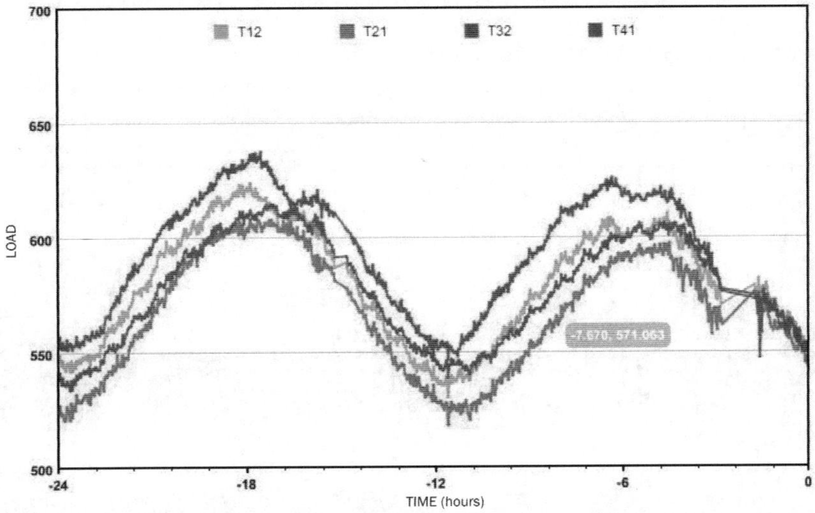

LAND | DEEPWATER | AEROSPACE | astrotechnology.com

ADVANCED DEEPWATER MONITORING SUMMARY

- Flow Lines (Temperature, Pressure, Hydrates)
- Risers, Tendons (Strain, Fatigue, Vibration)
- New Data Acquisition Overview
  - Calibration on Post-Installed Systems
  - Key Software Architecture Elements
- Observing Unique and Interesting Events
- Synthesizing Data into Actionable Information
  - Delivered to Key Decision Makers
- Future Activities Include Automation of Load Balancing and Abnormal Event Detection

LAND | DEEPWATER | AEROSPACE | astrotechnology.com

astro technology

CLEAR GULF JOINT INDUSTRY PROJECT (JIP)

- Collaboration between oil and gas industry, NASA and Astro Technology

PROPOSED PARTNERS INCLUDE:

ExxonMobil, Apache, ConocoPhillips, Chevron, Hess, Shell, Woodside, Anadarko Petroleum Corporation, Marathon Oil, PEMEX, Petrobras, Eni, bp, Total, Statoil

- Create cutting-edge techniques for managing production
- Develop environmental and safety systems for drilling and production
- Respond to challenges faced when working in remote and harsh environments
- 9 specific tasks

LAND | DEEPWATER | AEROSPACE | **astrotechnology.com**

astro technology

ADDITIONAL CONTENT

LAND | DEEPWATER | AEROSPACE | **astrotechnology.com**

astro technology

INSTRUMENTATION OF NASA'S ROBONAUT HAND

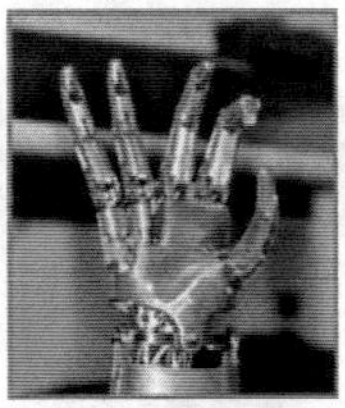

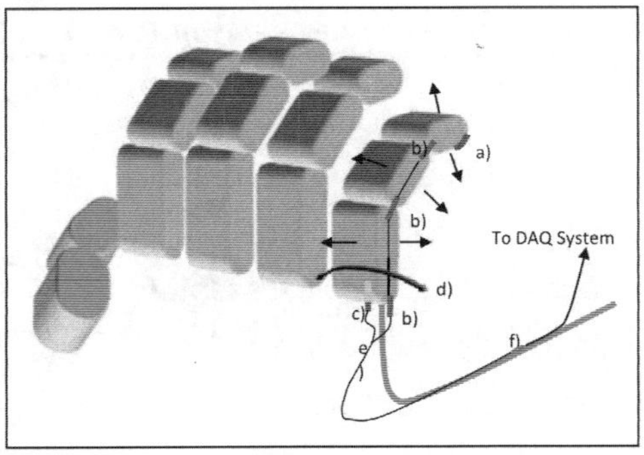

LAND | DEEPWATER | AEROSPACE | **astrotechnology.com**

astro technology

Space Shuttle Return-to-Flight

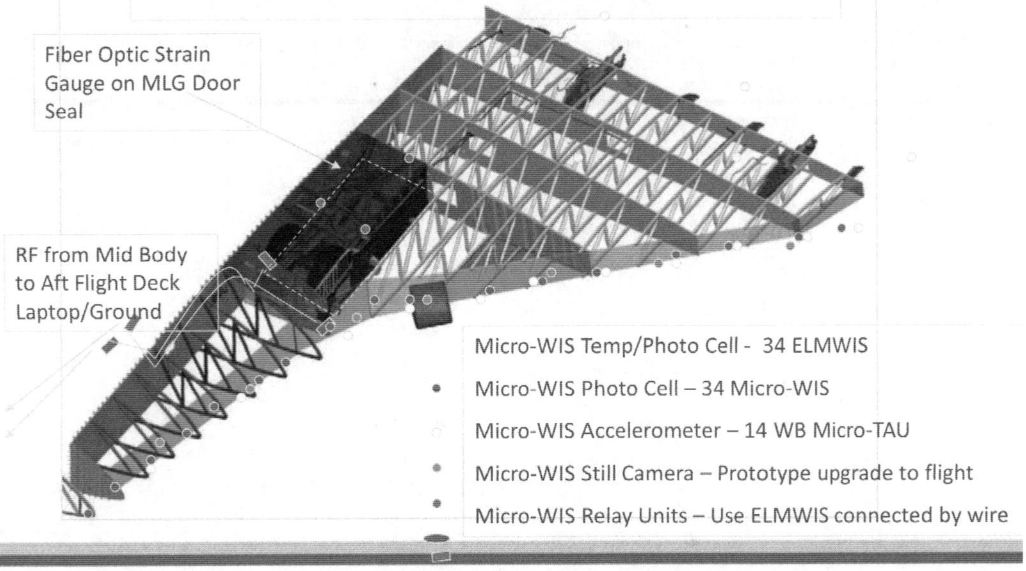

LAND | DEEPWATER | AEROSPACE | **astrotechnology.com**

astro technology

Solid Rocket Motor Test

- Apply fiber-optic sensor technology to solid rocket motors
  - In-situ sensors on new motors
  - Sensors bonded to the interior of existing motors
    - Fiber Bragg-gratings
    - Fabry Perot
- Measure mechanical and chemical properties during handling, storage, long-term aging, motor firing, case burst and damage assessment
- Provide tool for service life evaluation
- Early detection of possible structural failure

LAND | DEEPWATER | AEROSPACE | **astrotechnology.com**

astro technology

FIBER-OPTIC SENSOR ROCKET MOTOR TEST

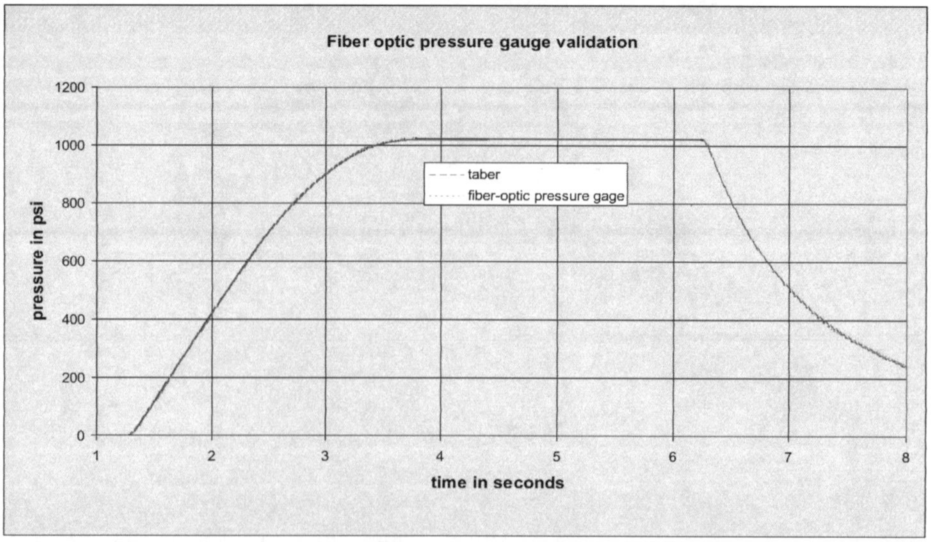

LAND | DEEPWATER | AEROSPACE | **astrotechnology.com**

FIBER-OPTIC SENSORS FOR CIVIL STRUCTURES

HARDENED SENSOR STATIONS

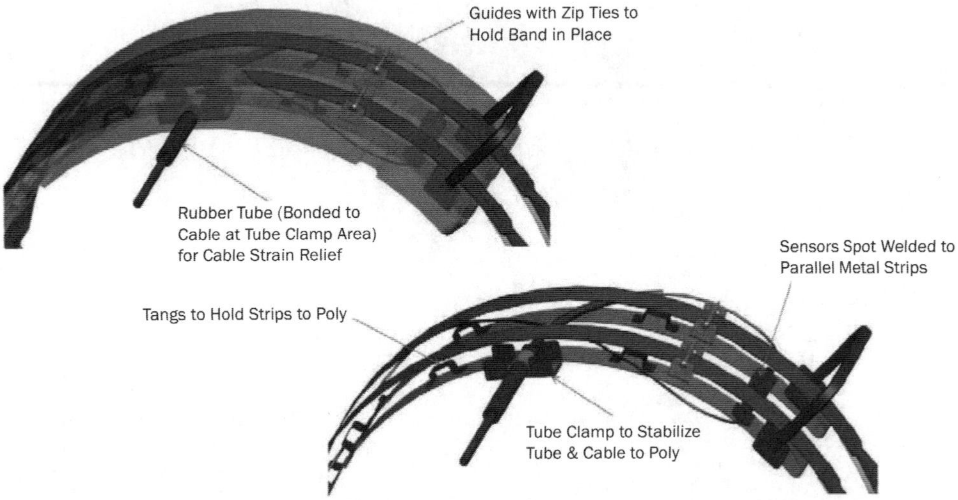

astro technology

OBSERVING WAVE ACTION

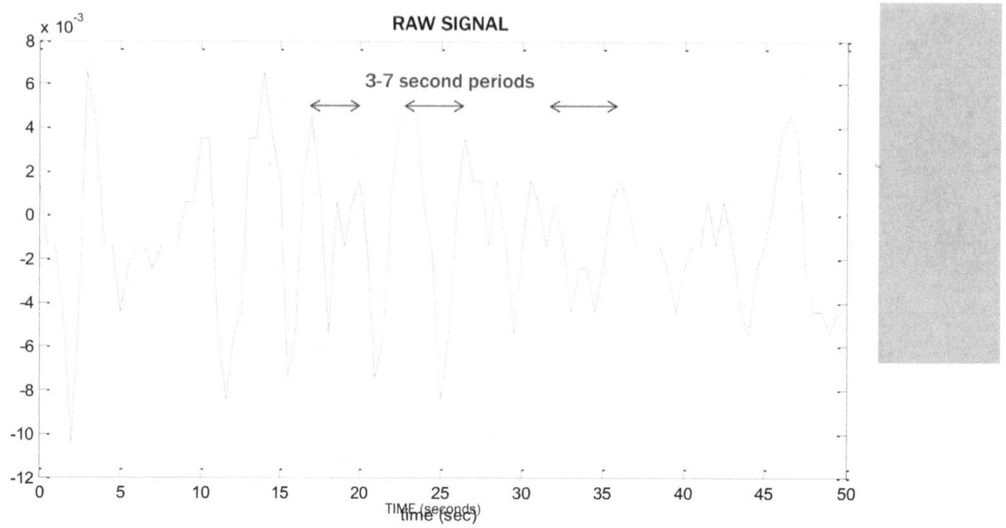

LAND | DEEPWATER | AEROSPACE | **astrotechnology.com**

astro technology

AMPLITUDE SPECTRUM

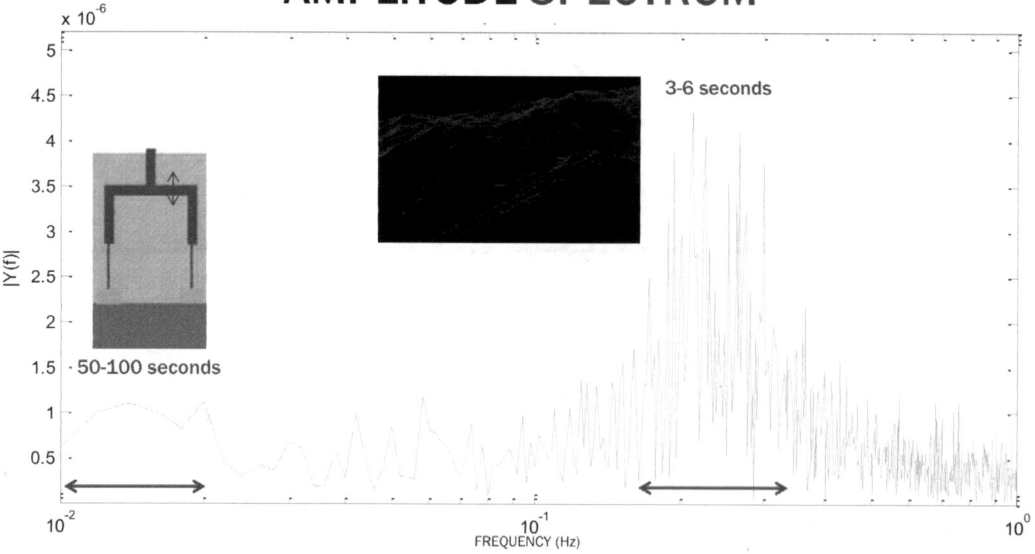

LAND | DEEPWATER | AEROSPACE | **astrotechnology.com**

352754 Characterization of Gas Hydrates in Turbulent Two-Phase Flow Environment Using CFD: Tracking Deposition, Aggregation and Dissociation Effects

Abstract:

Prediction and mitigation of gas hydrates continue to dominate the research in the oil and gas community owing to their ability to substantiate severe production losses and risk issues in many gas and oil fields. As the industry looks towards more reliable modeling for risk assessment and flow assurance, numerical tools continue to expand as a major asset in the research and development for the prediction and mitigation of gas hydrates.

In this paper, we establish a numerical framework to model the formation and transport of hydrate particles in an oil-water pipeline with finite concentration of dissolved methane in a horizontal flow line. Multiphase flow of dispersed water and gas in an oil dominated pipeline is captured using an Eulerian multi-fluid modeling method while accounting for the exchange of mass, momentum and energy between different phases. In particular, formation and transport of hydrate particles generated as a result of dispersed water droplets interaction with the dissolved methane gas, under thermodynamically favorable condition, is tracked using a population balance model. The nucleation process is modeled as a heterogeneous reaction, with intrinsic kinetics, between the water and gas phases leading to finite size hydrate particles. Subsequent growth, aggregation and deposition of the hydrate particle and its influence on flow modulation such as flow jamming and blockage are captured in an extensive fashion. Based on the incoming gas composition, the hydrate formation threshold temperature is established. The effective viscosity of the oil-water is modeled

The current study explores the effect of different incoming gas compositions (effective hydrate formation temperature), water cuts and pipe sizing on the hydrate formation rates and the resulting pressure gradients across the flow line under consideration. Contour plots of hydrate fraction and their size distribution along with velocity and pressure distribution provide a detailed understanding of the flow physics associated with the hydrate formation process.

349380 Assessing the Performance of a Kinetic Hydrate Inhibitor Using a Crystal Hydrate Inhibition (CGI) Method for a Gas and Gas-Condensate System

Abstract:

Gas hydrate formation in oil and gas pipelines is a complex process spanning space and time dimensions across multiple scales. At the micro/meso-scopic scales, (heterogeneous) nucleation and growth occurs. During nucleation, the first hydrate proto-nuclei (clusters) are formed, which, after some time 'induction time', can grow to produce detectable macroscopic crystals. Then, hydrate crystals may agglomerate, potentially forming a plug in the pipeline.

One of the methods to manage hydrate formation is the injection of Kinetic Hydrate Inhibitors (KHIs). The accepted understanding is that KHIs slow down or interfere with nucleation, hence extending the induction time. KHIs are normally evaluated by measuring induction times in the laboratory. However, quantifying induction times is problematic as nucleation is inherently stochastic, meaning measurements are often poorly repeatable with such studies time-consuming. Furthermore, it is often mistakenly believed that when a hydrate crystal has been detected in a KHI inhibited system, the KHI has 'failed'.

Recent research has proven that KHIs are not only nucleation delayers but also powerful crystal growth inhibitors. In view of this, a Crystal Growth Inhibition (CGI) method [1,2] for the reliable evaluation of KHIs has been previously developed. This method shows that KHIs induce a number of well-defined hydrate crystal growth inhibition (CGI) regions with different growth rates as function of subcooling. This new methodology provides a useful decision-making tool in order to manage the uncertainty of KHI performance during the different phases of real field production operations [3].

In this contribution, we present the results of an extensive experimental programme utilising the CGI method for the qualification of a commercial KHI that will be implemented in a greenfield development in the North Sea.

[1] Anderson et al. (2011) Development of a crystal growth inhibition based method for the evaluation of kinetic hydrate inhibitors, 7th Intl Conference on Gas Hydrates

[2] Anderson et al. (2011) Application of a new crystal growth inhibition based KHI evaluation method to commercial formulation assessment, 7th Intl Conference on Gas Hydrates

[3] Luna-Ortiz et al. (2013) Fast-track flow assurance design for kinetic hydrate inhibitors without laboratory testing: A case study, in 16th Intl Conference on Multiphase Production Technology

PROSYS™ Maximize Your Plant.

Operator Errors and What Can be Done to Minimize

Global Congress on Process Safety

March 31, 2014

10TH GLOBAL CONGRESS ON PROCESS SAFETY

PROSYS™ Maximize Your Plant.™

10TH GLOBAL CONGRESS ON PROCESS SAFETY

Presenter

Tom Nolan

- **Graduated from Ohio University with B.S. in Chemical Engineering**
- **24 years experience in chemical process industry in a variety of roles**
- **MAIC and DFSS Six Sigma Black Belt**

PROSYS™ Maximize Your Plant.™

The Cost of Errors

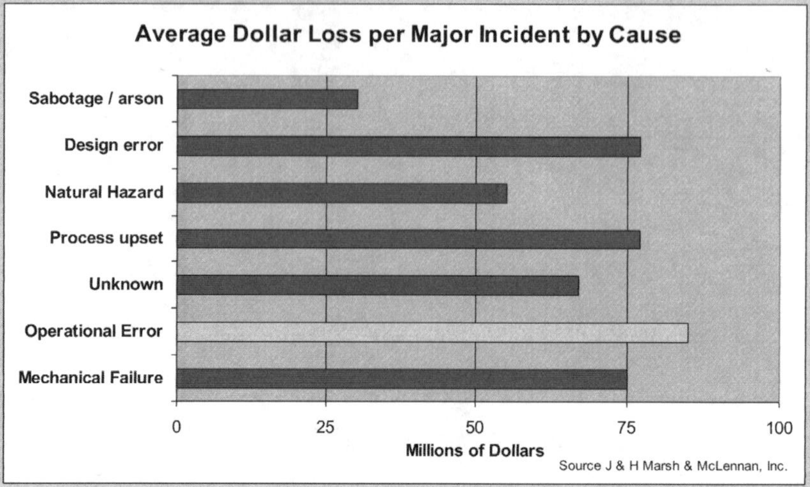

PROSYS™ Maximize Your Plant.™

The Cost of Operator Errors

- **ASM estimates total loss due to operator error is $8B per year**
- **Chemical Safety Topical Committee - average of one chemical incident per day - cost of over $2 million per incident to <u>comply with requirements</u>**
- **Errors cause 42% of unscheduled shutdowns**
- **70% of process incidents occur during start-up or shutdown**

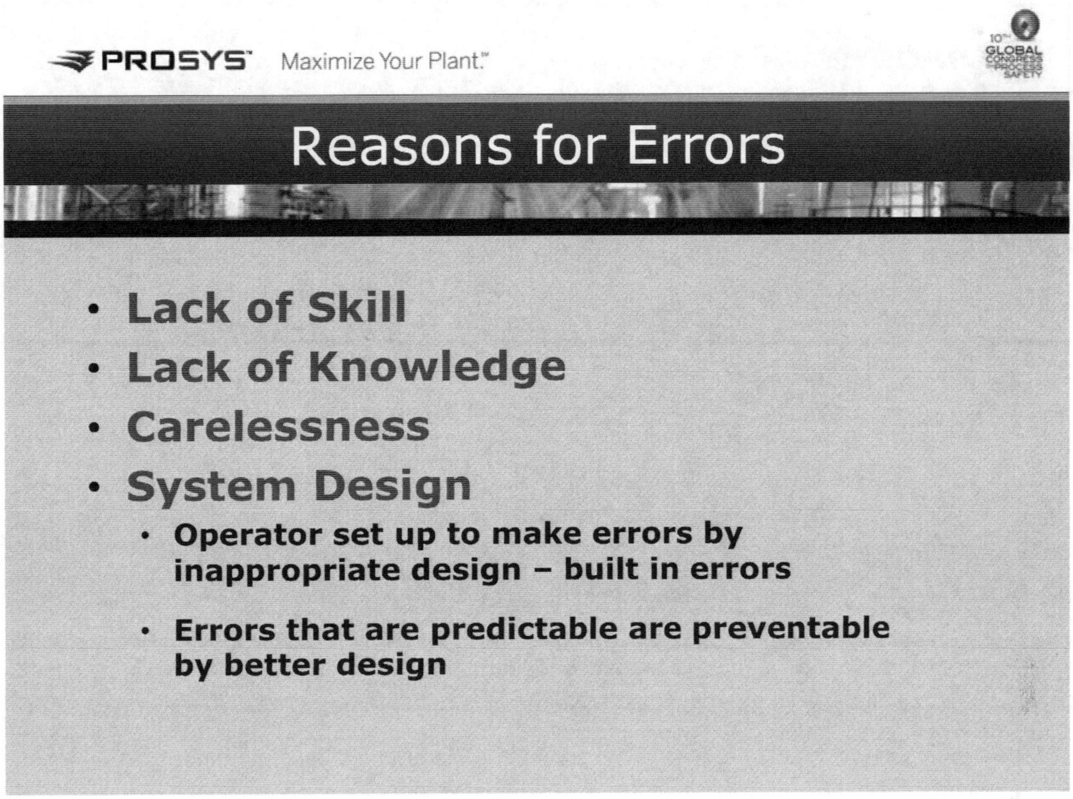
PROSYS Maximize Your Plant.
10th GLOBAL CONGRESS ON PROCESS SAFETY
Reasons for Errors
• Lack of Skill
• Lack of Knowledge
• Carelessness
• System Design
• Operator set up to make errors by inappropriate design – built in errors
• Errors that are predictable are preventable by better design

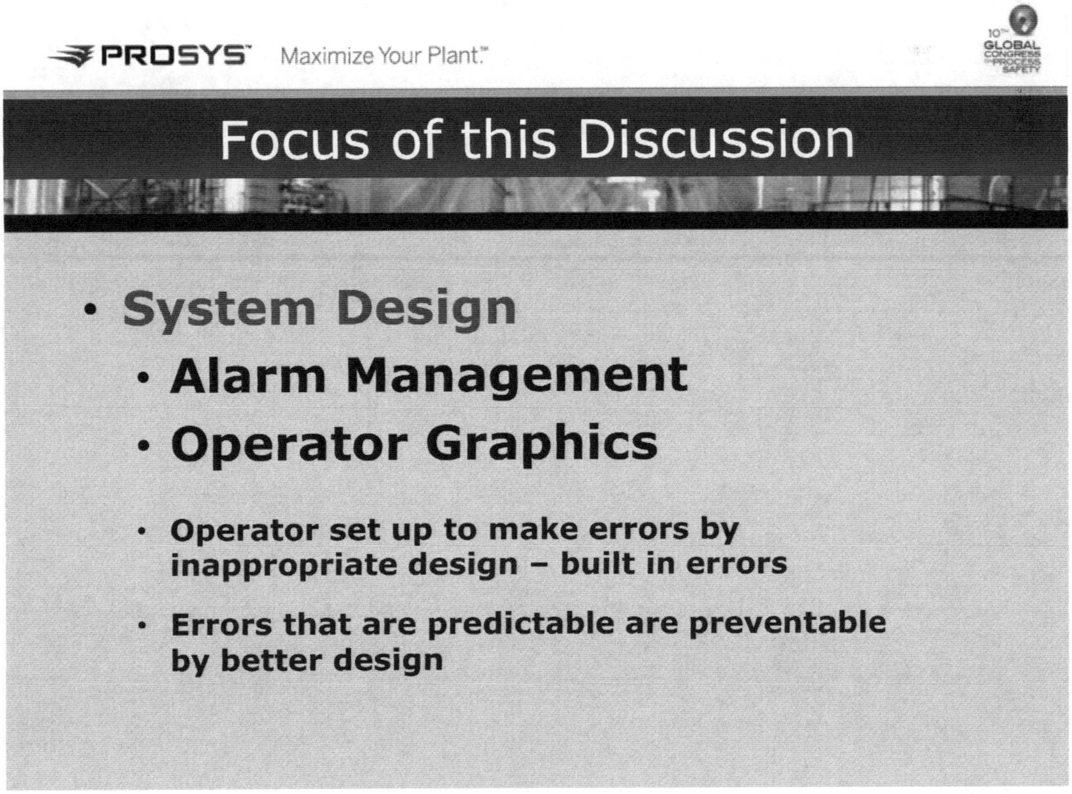
PROSYS Maximize Your Plant.
10th GLOBAL CONGRESS ON PROCESS SAFETY
Focus of this Discussion
• System Design
• Alarm Management
• Operator Graphics
• Operator set up to make errors by inappropriate design – built in errors
• Errors that are predictable are preventable by better design

PROSYS Maximize Your Plant™

10th GLOBAL CONGRESS ON PROCESS SAFETY

Alarm Floods

In a number of industrial incidents, alarm floods were identified as a significant contributing cause to the incident...

As found by EEMUA in 1999 and CSB

Alarm Flood defined by ISA 18.2 as -
"10 or more annunciated alarms in any 10 minute period per operator"

PROSYS Maximize Your Plant™

10th GLOBAL CONGRESS ON PROCESS SAFETY

Why do Alarm Floods Occur?

One reason is not providing dynamic alarm management

- **Alarms need to indicate abnormal situations that require operator action**
- **Processes do not operate in one state**
- **What is normal vs abnormal changes with state**
- **Alarms are typically configured for run therefore many alarms are triggered upon a Change of State: Run to Shutdown**
- **Many of the alarms are not applicable or actionable for the new state impleading the operators ability to act quickly on what is important**

PROSYS Maximize Your Plant.™

10th GLOBAL CONGRESS ON PROCESS SAFETY

What is Impacted by Alarm Floods?

- Product quality
- Operability or profitability of the process
- Loss of equipment
- Loss of containment – environmental releases
- Injury and loss of life in plant or community

PROSYS Maximize Your Plant.™

10th GLOBAL CONGRESS ON PROCESS SAFETY

What Makes Alarm Floods so Dangerous?

Can be a problem for three reasons:

- A deluge of alarms can cause critical alarms to be missed
- Floods can be a significant distraction when dealing with process upsets
- Can be an indicator of larger systemic safety issues

PROSYS Maximize Your Plant.™

10th GLOBAL CONGRESS ON PROCESS SAFETY

Impacting Alarm Management Design

- Alarm rationalization is not a process to eliminate alarms - its about quality
- Good rationalization will add alarms when appropriate
- Typical (Static) rationalization is only for run mode
- Dynamic rationalization considers all plant modes

PROSYS Maximize Your Plant.™

10th GLOBAL CONGRESS ON PROCESS SAFETY

Dynamic Alarm Management

- Dynamic rationalization does everything a static rationalization does plus asks "when" for each alarm
- Requires dynamic software to make changes based upon operating mode of the plant
- Eliminates redundant alarms and lowers operator loading during transitions
- Only alarms what is abnormal and actionable for the given state

PROSYS Maximize Your Plant.™

10th GLOBAL CONGRESS on PROCESS SAFETY

Actual Performance Metrics vs ISA 18.2

Type	Avg Alarm Rate/hr	Low/Hi of Avg Rate/hr	Peak Alarm Rate/hr	Low/Hi of Peak Rate/hr	% Time in Flood	Standing Alarms >24 hrs
Before Rationalization	30	Low=6.2 Hi=61	638	Low=152 Hi=2402	17.2%	9
ISA 18.2 Target Metrics	6	---	<60	---	<1%	<5
After Dynamic Alarm Management	2	Low=0.09 Hi=2.5	25	Low=13 Hi=42	0.25%	4

PROSYS Maximize Your Plant.™

10th GLOBAL CONGRESS on PROCESS SAFETY

Results After Dynamic Alarm Management

Annunciated Alarms Per Hour (Monthly) **Area: G1-G5-G6** **June 2013**

Average Counts | Peak Counts | Total Counts

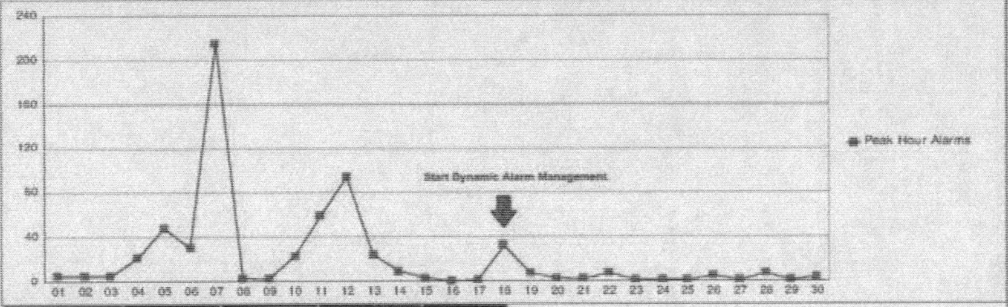

Day	Min Hour Alarms	Avg Hour Alarms	Peak Hour Alarms	Total Alarms
01-Jun-2013	0	0.3	5	6
02-Jun-2013	0	0.4	5	9
03-Jun-2013	0	0.5	5	13
04-Jun-2013	0	1.1	21	27
05-Jun-2013	0	13.3	48	319
06-Jun-2013	0	3.6	30	86
07-Jun-2013	0	11.9	215	285
08-Jun-2013	0	0.4	3	10
09-Jun-2013	0	0.3	2	7
10-Jun-2013	0	1.5	23	35
11-Jun-2013	0	4.7	59	112
12-Jun-2013	0	11.9	94	285
13-Jun-2013	0	3.8	24	90
14-Jun-2013	0	1.3	9	32
15-Jun-2013	0	0.3	3	8

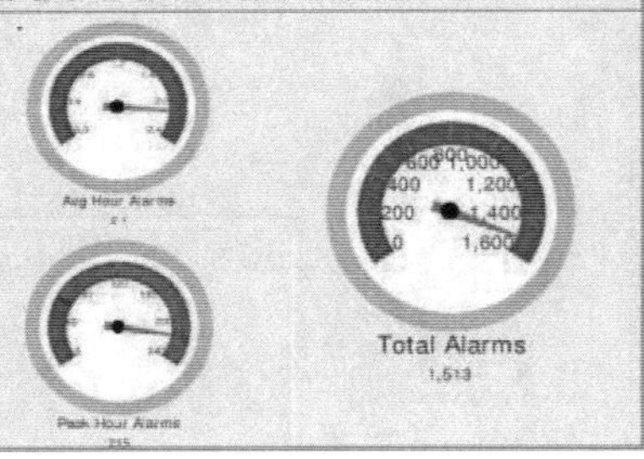

PROSYS™ Maximize Your Plant.™

10th GLOBAL CONGRESS ON PROCESS SAFETY

Operator Graphics – The Risk of Changing Focus

- **Distraction of changing focus from process graphic to faceplate window**
- **Added workload from managing multiple open windows**
- **Increased probability of errors when changing values for un-intended tag with multiple faceplates open**
- **All of above become multiplied and more complex when process is transitioning from one state to another**

PROSYS™ Maximize Your Plant.™

10th GLOBAL CONGRESS ON PROCESS SAFETY

Faceplate Operation

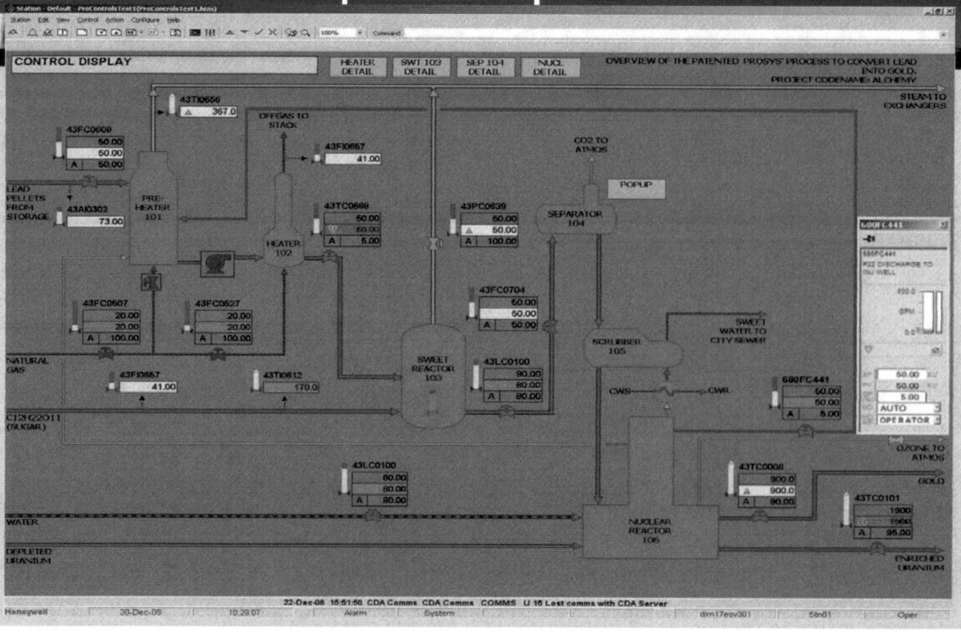

Potential for Errors

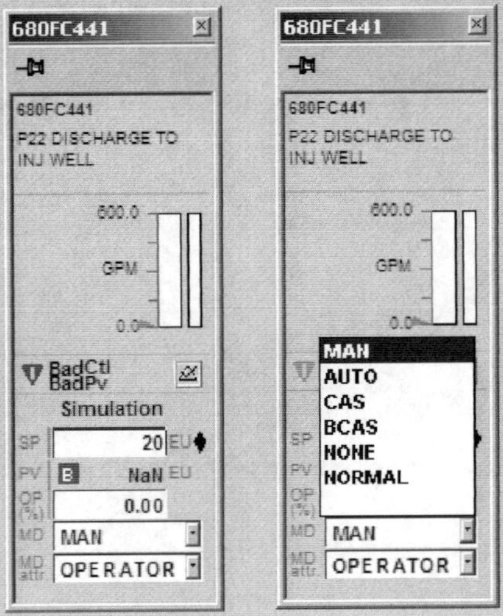

Faceplate Design -can introduce additional errors

- Mode drop down list covers SP, PV and OP values
- Operator may select wrong Mode if SP, PV and OP values are hidden from view
- Mode list offers more options than necessary

PROSYS™ Maximize Your Plant.™
10TH GLOBAL CONGRESS ON PROCESS SAFETY

Direct Entry Operation

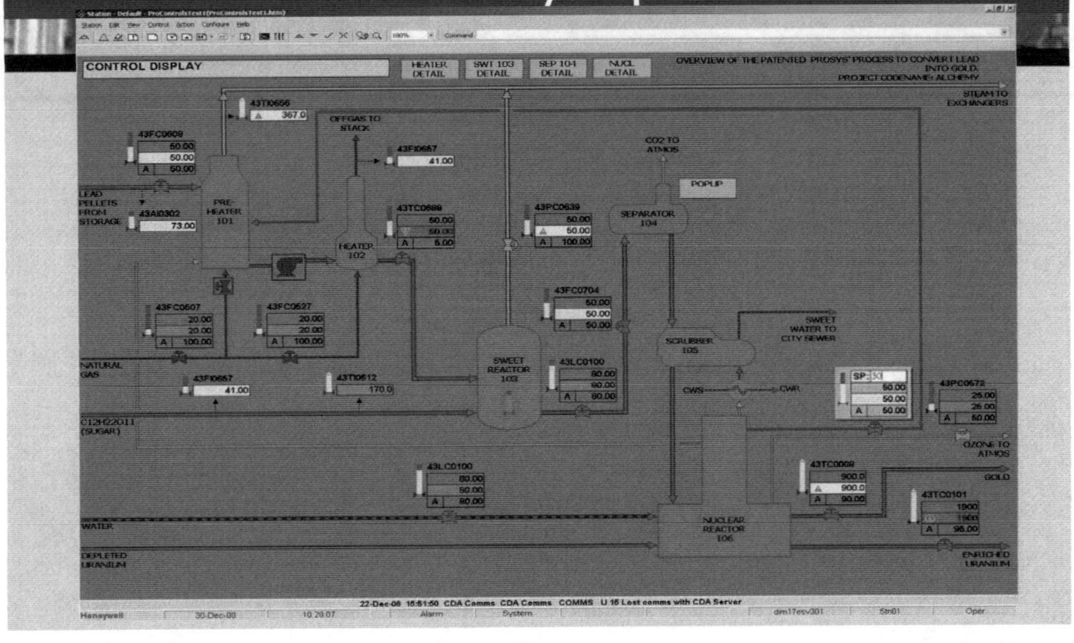

Direct Entry Fields

Values can be entered by:
- Select point
- Typing value via keypad
- Pressing enter

Key Factors:
- By maintaining focus of points after value change, it improves safety for input errors, multiple clicking functions and is more time efficient
- Easily recognizable operator enterable fields

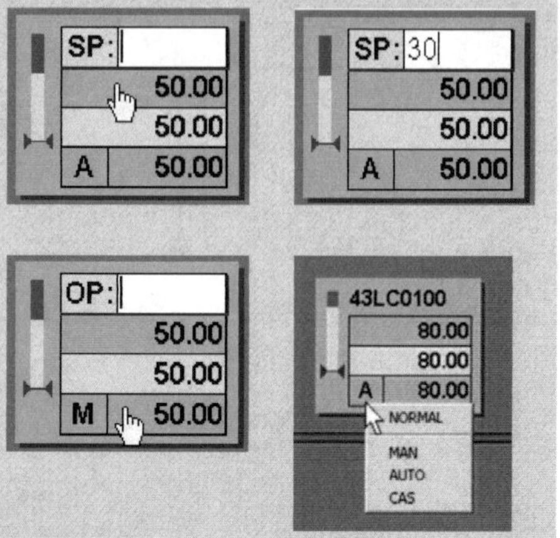

Evaluation of Methods

Comparison of Faceplate to Direct Entry
- **Keystroke Level Analysis**
- **Operator Loading Analysis**
- **Risk Analysis**

PROSYS Maximize Your Plant.™

10th GLOBAL CONGRESS on PROCESS SAFETY

Keystroke Level Model

Symbol	Time (s)	Description
K	0.28	Keystroke
P	1.1	Point to object
BB	0.2	Click on object
H	0.4	Home hands on keyboard or mouse
M	1.2+	Mental act or routine thinking

PROSYS Maximize Your Plant.™

10th GLOBAL CONGRESS on PROCESS SAFETY

Estimate Execution Time Analysis

- **Evaluates execution time by an operator comparing faceplates to direct entry for SP/OP and Mode Changes**

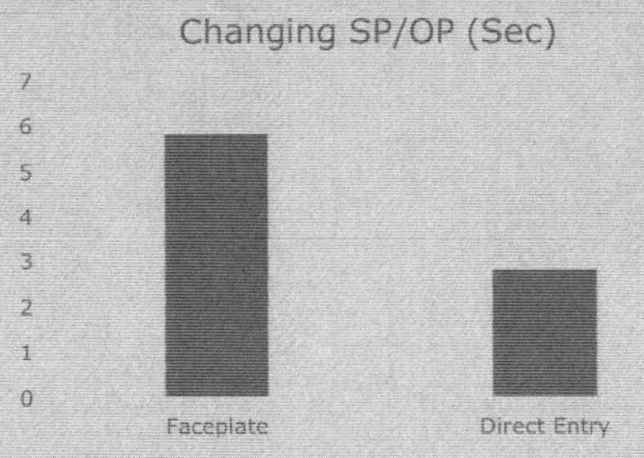

- **50 % Reduction**

PROSYS Maximize Your Plant™

10th GLOBAL CONGRESS ON PROCESS SAFETY

Estimated Operator Loading Analysis

- **Evaluates Mental and Physical Load on Operator**
  - Tasks that require very little thought are removed – such as (BB), (H)

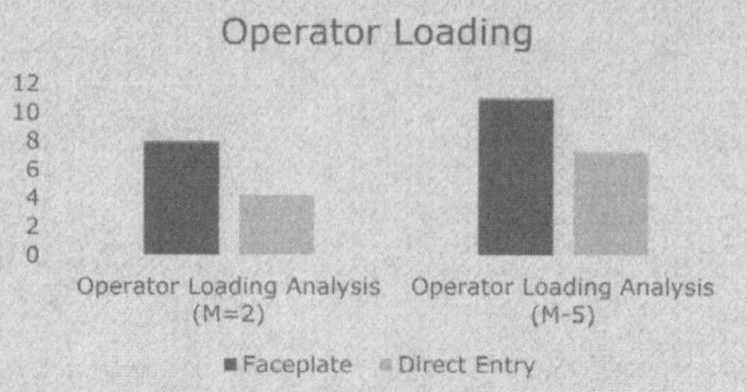

  - Mental and Physical tasks are weighted based on the amount of fatigue they cause – assigned a weight of M=2.0 to 5.0

PROSYS Maximize Your Plant™

10th GLOBAL CONGRESS ON PROCESS SAFETY

Risk Analysis

- **Identifies actions where operator entry errors can occur**

PROSYS Maximize Your Plant.™

10th GLOBAL CONGRESS on PROCESS SAFETY

Risk Analysis

Faceplate operation KLM for changing SP/OP	Potential Error
Move to shape (P)	Select wrong point
Click on shape (BB)	
Move to faceplate (P)	Point in faceplate is previous point
Click on SP/OP field (BB)	Change wrong parameter
Move hands to keyboard (H)	
Type in value and press enter (4K)	Mistype value and press enter
Move hands to mouse (H)	
Move to close faceplate (P)	
Click to close faceplate (BB)	
Potential error count	**4**

PROSYS Maximize Your Plant.™

10th GLOBAL CONGRESS on PROCESS SAFETY

Risk Analysis

Direct Entry operation KLM for changing SP/OP	Potential Error
Move to SP/OP (P)	Select wrong parameter or point
Click on SP/OP field (BB)	
Move hands to keyboard (H)	
Type in value and press enter (4K)	Mistype value and press enter
Potential error count	**2**

PROSYS Maximize Your Plant.™

10th GLOBAL CONGRESS ON PROCESS SAFETY

Direct Entry Vs Faceplate

Summary of Benefits

- **Execution Time SP/OP Changes - 51.5% less**
- **Execution Time for Mode Changes – 50% less**
- **Operator Loading – 33 % to 47% less**
- **Risk of Errors – 50% less**

PROSYS Maximize Your Plant.™

10th GLOBAL CONGRESS ON PROCESS SAFETY

Conclusion

- **Many industrial errors are a result of operators using control systems with flawed designs**
  - Poor design and performance by alarm management distracts operators and/or occludes critical alarms
  - Operator graphics with multiple open faceplates can cause a change intended for one controller to be entered into the faceplate of another
- **The two mechanisms listed above are predictable and therefore preventable through better design available today**

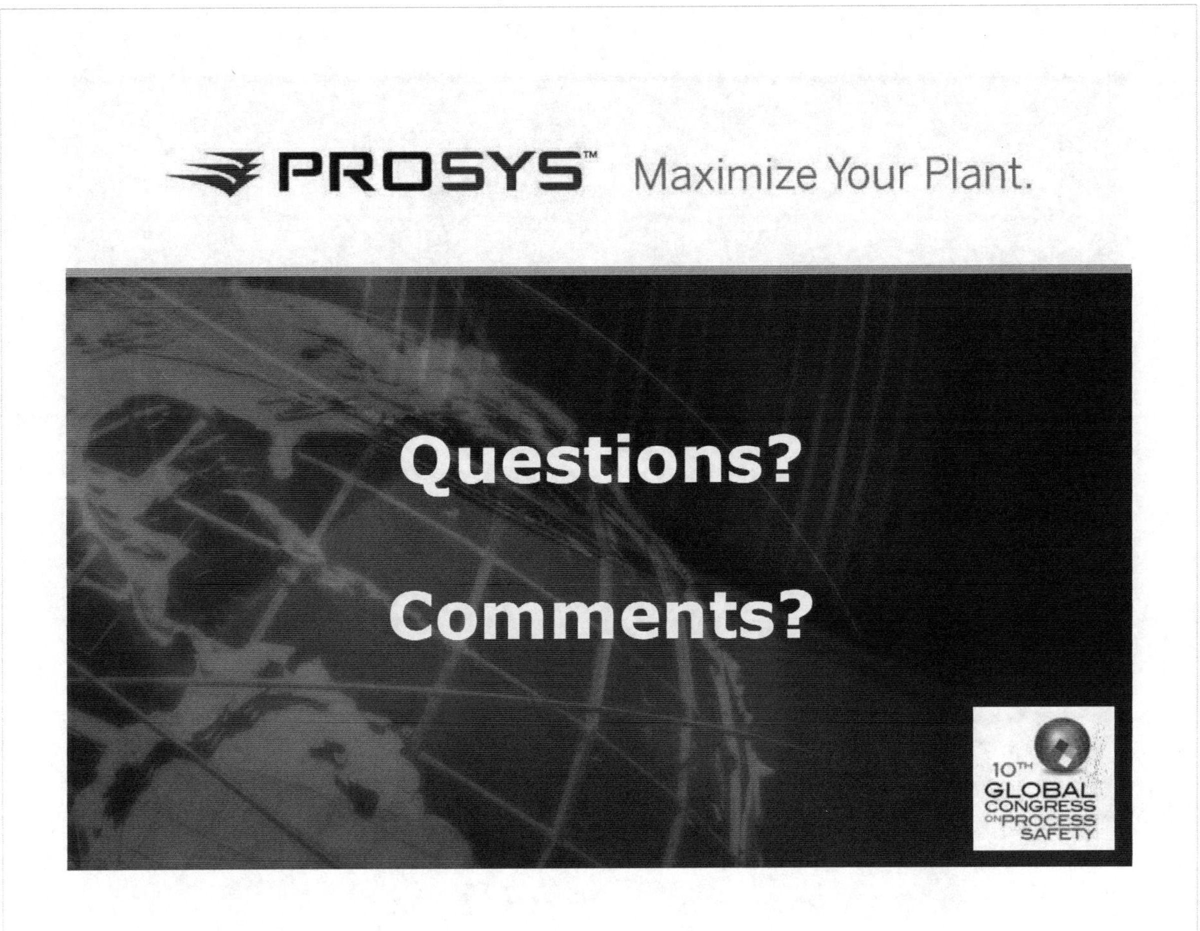
PROSYS™ Maximize Your Plant.
Questions?
Comments?
10TH GLOBAL CONGRESS ON PROCESS SAFETY

Problem Statement

visium KMS
Process Safety Management Solution

- Any organisation can improve by leveraging knowledge and lessons learned.
- Knowledge can be externally or hidden locally, in a global enterprise, where lessons learned are difficult to share beyond their initial impact.
- There is so much data, information and knowledge (opportunities) coming at your company that can make you stronger or make you look like your not paying attention.
- Not paying attention or managing the available knowledge, can sink your ship… nobody wants that!

Expert Insight Delivered 2 OSyS
Part of the Rolls-Royce Group

What is Knowledge Management?

visium KMS
Process Safety Management Solution

- Knowledge management is the concept in which an enterprise consciously and comprehensively gathers, organizes, shares, and analyses its knowledge to further its aims.
- Turning data into information is nice.
- Turning Information into Knowledge is good!
- Turning Knowledge into Wisdom is great!
- But something is missing…
- The CAPA (doing) loop…

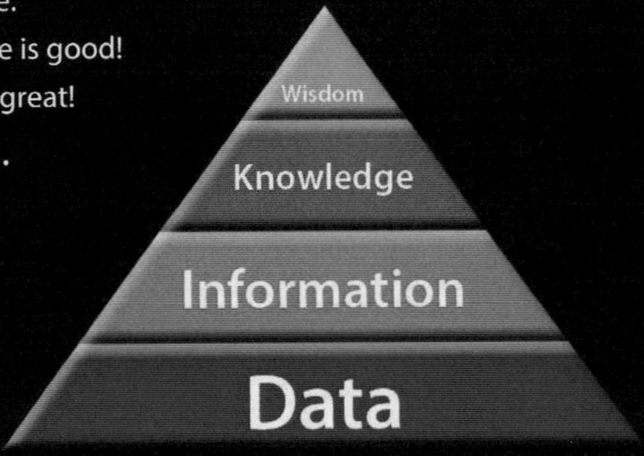

Expert Insight Delivered

3

Osys
Part of the Rolls-Royce Group

What we need is this…The HUAA MS

visium KMS
Process Safety Management Solution

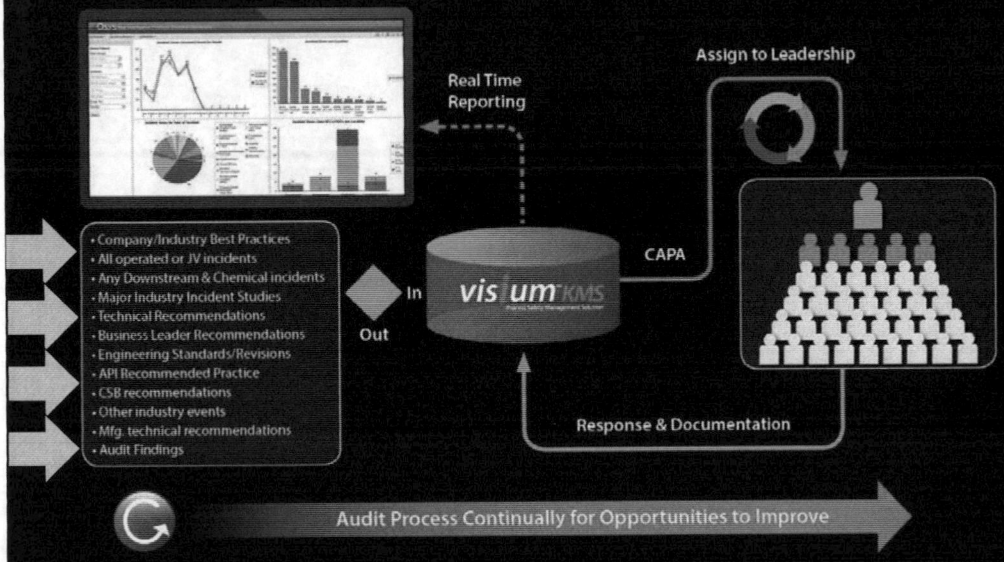

Expert Insight Delivered

4

Osys
Part of the Rolls-Royce Group

What is the HUAA Management System…

visium KMS
Process Safety Management Solution

HUAA – Heard, Understood, Acknowledged, Actioned.

1. Opportunities identified (heard) by your experts,
2. Captured by an electronic system,
3. Expertly Reviewed (understood),
4. Accepted or rejected (acknowledged) and
5. Assigned (actioned) to Sr. Managers for resolution and tracked to closure.

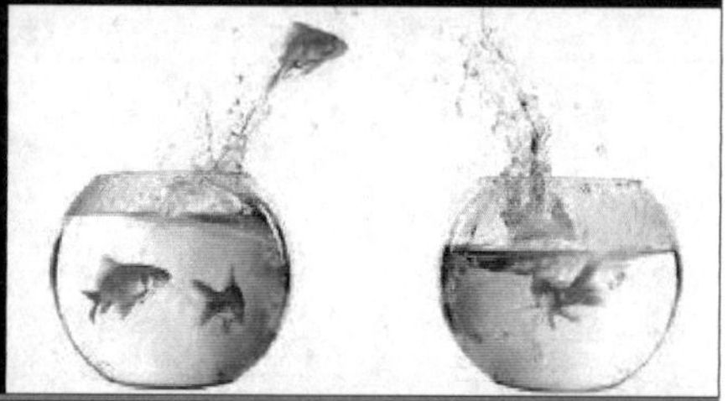

Expert Insight Delivered 5

OSyS
Part of the Rolls-Royce Group

HUAA – Management Systems - Detail

visium KMS
Process Safety Management Solution

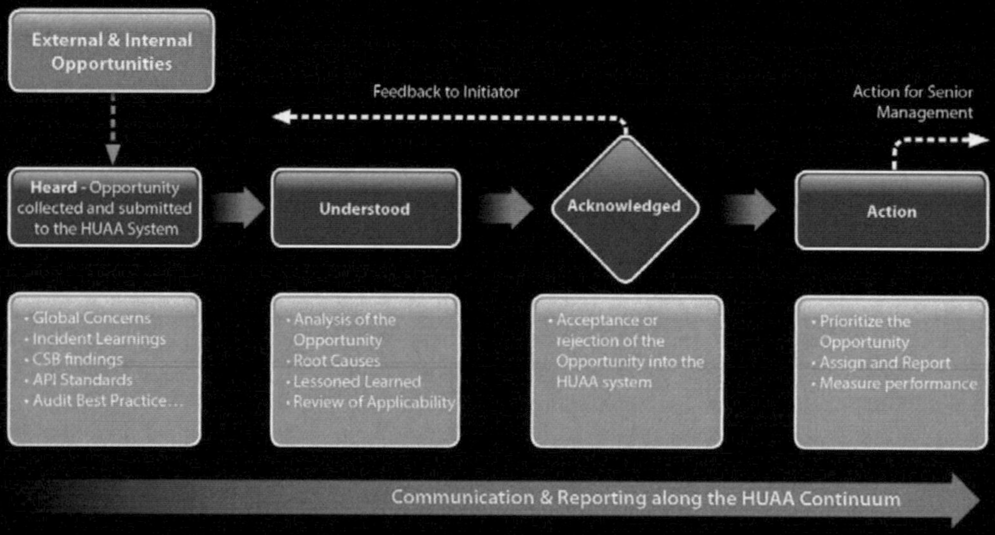

Expert Insight Delivered 6

OSyS
Part of the Rolls-Royce Group

HUAA Results

visium KMS
Process Safety Management Solution

- Efficient Collection - Important Opportunities are identified and collected from many sources (internal and external), and entered into the HUAA management system.
- Expert Vetting - These opportunities are vetted, parsed and resolved for inclusion or exclusion; documenting & communicating both.
- Accountable Assignment - Included opportunities are systematically assigned to Sr. Leaders and managed across the enterprise at SVP and C levels.
- Action - Improvement will be assured through the accountable actions of these leaders.
- Results - The organization will gain competitive advantage by active listening, learning from their experience, as well as the experience of others, and by taking ACTION!

Expert Insight Delivered 7

Osys
Part of the Rolls-Royce Group

Sharing is not enough to keep your ship afloat!

visium KMS
Process Safety Management Solution

Expert Insight Delivered 8

Osys
Part of the Rolls-Royce Group

352631 Autocad - Smart Grid – Agriculture - Architecture and the United States Government

Abstract:

The computational methods, process control, process design, process safety, public policy of the smart chip for agriculture industry, and architecture industry in the United States utilizing the computational methods of AutoCAD in the structural development stages of the smart grid for device/component/equipment and geographical development and public policy for grant funding.

352747 Production, Characterization and Catalytic Studies of Biobased Carbon Materials

Abstract:

Carbon-based nano materials were produced from wood chips, lignin and bio-char. The influence of reaction conditions on the production of nano-carbons was investigated. Carbon-based nanomaterials were characterized using scanning electron microscopy (SEM), transmission electron microscopy (TEM), high resolution transmission electron microscopy (HRTEM), and X-ray diffraction (XRD). The catalytic performance of these nanomaterials as syngas cleaning adsorbents and chemical conversion catalysts for bio-fuel processes has been evaluated.

347835 Uncertainty in Sour Gas Viscosities Estimation, What Is the Effect on Your Reservoir Inflow and Tubing Performance

Abstract:

In light of the need to find an applicable, consistent and promising correlation to accurately calculating sour gases viscosity under several reservoir conditions, this study aims to compare the existing correlations and other industry standard methods, check their accuracy and range of applicability for sour/acidic gases. Moreover, investigating the impact of uncertainties in estimating sour gas viscosities on reservoir flow performance relationship (IPR), reserves and resources estimations, tubing performance, pipeline and facilities design, the design of compressors and more importantly the need to find such correlation to be used in further advancements of reservoir simulation.

354052 Chromonic Nanocarriers for Chemotherapeutics: Size Distribution and Control Release Studies

Abstract:

Conventional chemotherapy fails to release the drug in the controlled manner; rather they release the drug instantly due to which re-administration is required as therapeutic index falls. Moreover, they target both the cancer cells and non-cancer cells which make the treatment of the cancer cells highly ineffective due to excessive toxicities. Recently, the nanoparticle mediated drug delivery for chemotherapeutics has gained much attention. The present study reports the use of a novel class of materials – *Chromonics* in cancer drug delivery. Folic acid as a chromonic material has been explored to develop nanocarrier for sustained release of chemotherapeutic drugs *in-vitro*. With the help of a biocompatible polymer as a tool, folate nano-domains were formed from self-assembled liquid crystalline folate solutions. These nano-domains were crosslinked to form a stable nanoparticle system. The effect of different factors like crosslinking agent, relative polymer and folate concentration, on the size distribution was investigated in detail. Based on the observations, we have reported that not only the synthesis but also the size of the developed nanoparticles can be controlled by varying the folate: HPMC weight ratios. The present study then explored the release of drugs from these nanocarriers. We have investigated the release of folic acid as a model drug from these nanoparticles as well as release of methotrexate and cytarabine as an anticancer drug for the period of 60 days. It was observed that release of drug was nanoparticle's size-dependent, release medium dependent, pH of the release medium dependent, crosslinking agent and its concentration dependent. The significant part of the current study is the synthesis of chromonic nanoparticles, how the size can be controlled, how controlled release can be obtained as well as how the release profiles can be modified depending on the dosage forms. Therefore, we can say that folate nanoparticles can be a potent nanocarrier for the sustained release of cancer drugs.

Double Containment Piping Systems: A Fail-Safe, Leak-Free Solution

IPEX
Committed to Excellence

Darin Johnson, P.Eng.
Applications Engineer
IPEX USA LLC

AIChE – 2014 Spring Meeting
New Orleans, LA

Monday, March 31, 2014

Why Double Containment Piping Systems?

IPEX
Committed to Excellence

- December 1988 – law change within the ***Resource, Conversation, and Recovery Act*** in the United States regarding underground piping.
- U.S. Environmental Protection Agency (EPA) requirement: Code of Federal Regulations (CFR), Title 40, Part 280.
  - Petroleum Products
  - Hazardous Substances (over 3200, CFR Title 40 Part 302.4)
  - Hazardous Wastes

Why Double Containment Piping Systems?

- Buried tank - all buried piping connected to it must be double contained.
- Above ground tank – associated buried piping consists of 10% or more of the tank volume, it also must be double contained.

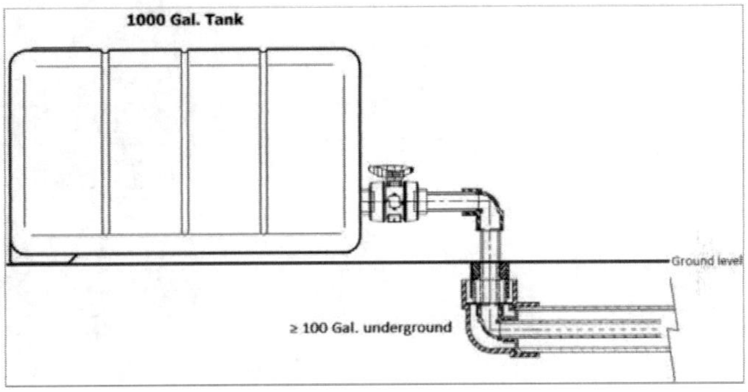

Why Double Containment Piping Systems?

- If the piping is less than 10% of the volume of the aboveground tank, and you choose not to double contain the buried piping, ***<u>responsibility for any clean-up or remediation resulting from a release will still fall on the Client.</u>***

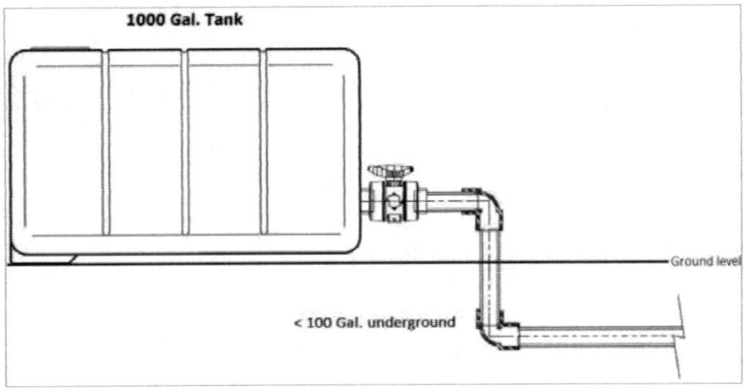

Why Double Containment Piping Systems?

- **Above Ground Piping Systems**
  - Different jurisdictions have to ability to require chemicals be double contained.

Nomenclature of Double Containment Piping

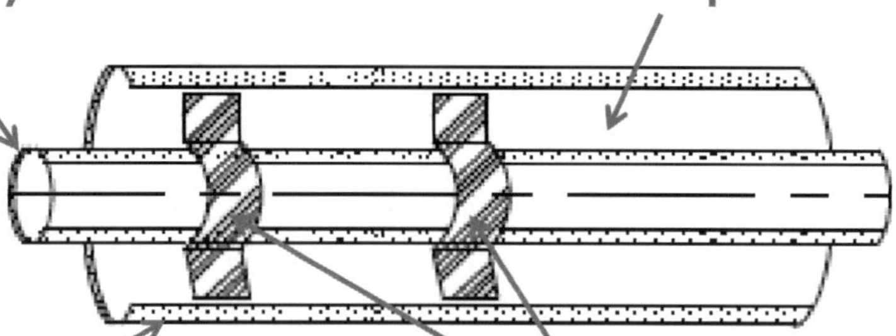

Double Containment Piping Systems: Material Combinations

SYSTEM Carrier	Containment	Carrier SIZE inches
PVC 80	PVC 80	1/2 - 12
PVC 80	PVC 40	1/2 - 12
PVC 40	PVC 40	1/2 - 12
PVC DWV	PVC DWV	1-1/2 - 12
PVC 80	CS SCH 10	1/2 - 12
CPVC 80	CPVC 80	1/2 - 8
CPVC 80	PVC 80	1/2 - 12
CPVC 80	PVC 40	1/2 - 12
CPVC 80	CS SCH 10	1/2 - 12
NAT PP 80	PVC 80	1/2 - 2
NAT PP 80	PVC 40	1/2 - 2
RED PVDF 80	CPVC 80	1/2 - 2
RED PVDF 80	PVC 80	1/2 - 2
NAT PVDF 80	PVC 80	1/2 - 2
ENFUSION PP	PVC DWV	1/2 - 6
S40 A53B CS BW	PVC 80	1/2 - 8
S40 A106B CS BW	PVC 80	1/2 - 8

SYSTEM Carrier	Containment	Carrier SIZE inches
S40 A53 BLK SW	FRP	1/2 - 2
S40 A53 BLK SW	PVC 80	1/2 - 2
S80 A53B CS SW	PVC 80	1/2 - 2
S10 T304L SS BW	PVC 80	1/2 - 8
S10 T304L SS BW	PVC 40	1/2 - 8
S10 TO304L SS BW	CPVC 80	1/2 - 8
S10 T304L SS BW	FRP	1/2 - 12
S10 T304L SS BW	S10 T304L	1/2 - 12
S40 T316L SS BW	PVC 80	1/2 - 8
S40 T316L SS BW	CPVC 80	1/2 - 8
S40 T316L SS BW	FRP	1/2 - 12
S40 T304L SS SW	PVC 80	1/2 - 2
S40 T304L SS SW	PVC 40	1/2 - 2
S40 T304L SS SW	CPVC 80	1/2 - 2
S40 T304L SS SW	FRP	1/2 - 2
S10 T316L SS SW	PVC 80	1/2 - 2
S10 T316L SS SW	CPVC 80	1/2 - 2

Double Containment Piping Systems: Design Considerations

- **Material Selection**
  - Primary Pipe
    - Chemical Compatibility
    - Pressure Capabilities
    - Thermal Properties
  - Secondary Pipe
    - Chemical Compatibility
    - Loading (Static & Live) for buried applications
    - Hanging requirements for above ground
    - Leak detection requirements
    - UV requirements

Double Containment Piping Systems: Design Considerations

- **Thermal Expansion & Contraction**
  - Two basic thermal relationships may be present:
    - When only one of the two systems experiences thermal expansion/contraction.
    - When both systems experience different magnitudes of thermal expansion/contraction.
  - The Floating System
  - The Restrained System

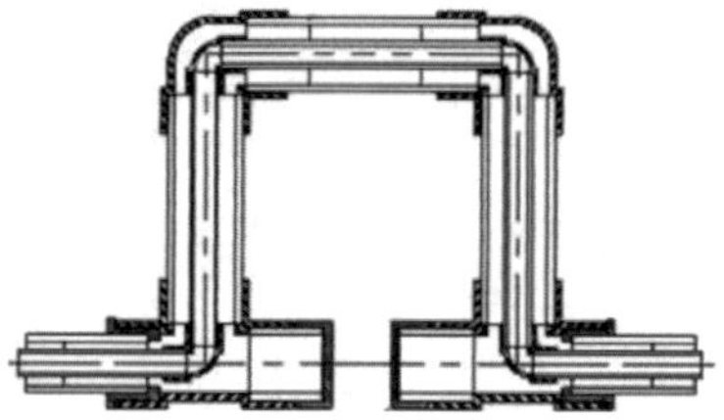

Double Containment Piping Systems: Leak Detection Requirements

- EPA requirement: (CFR), Title 40, Part 280
  - Pressurized systems require electronic leak detection that must be checked once per year.
  - Drainage systems (gravity flow) require visual leak detection with monthly manual inspection for a media leak.

Double Containment Piping Systems: Leak Detection Requirements

- The following forms of leak detection are acceptable as per regulations
  - Visual (drainage applications only)
  - Electronic low point probes
  - Continuous cable

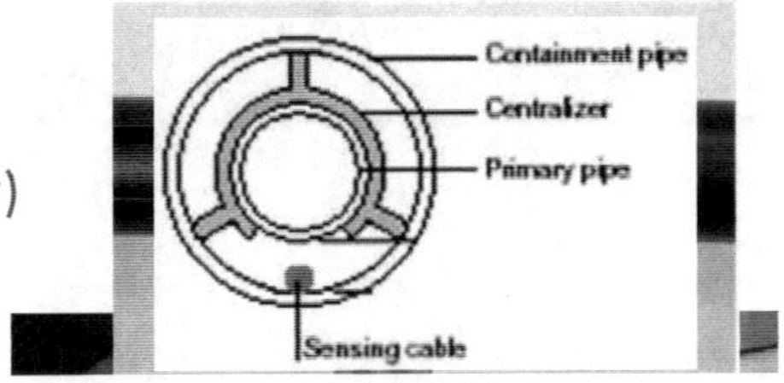

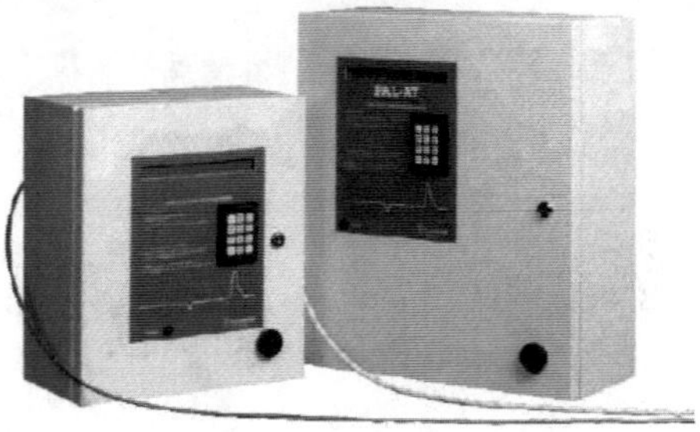

Double Containment Piping Systems: Applications

- Underground Applications
- WTP & WWTP
- Oil & Gas
- Hospitals
- Museums
- Food and Beverage
- Pharmaceuticals
- Laboratories
- Data Centers

IPEX
Committed to Excellence

Double Containment Piping Systems: A Fail-Safe, Leak-Free Solution

Questions?

Thank you!

DTP Process

"Now Ready to License"

JGC Corporation
Mitsuo Morita

2014

JGC CORPORATION | MITSUBISHI CHEMICAL

What is DTP Process ?

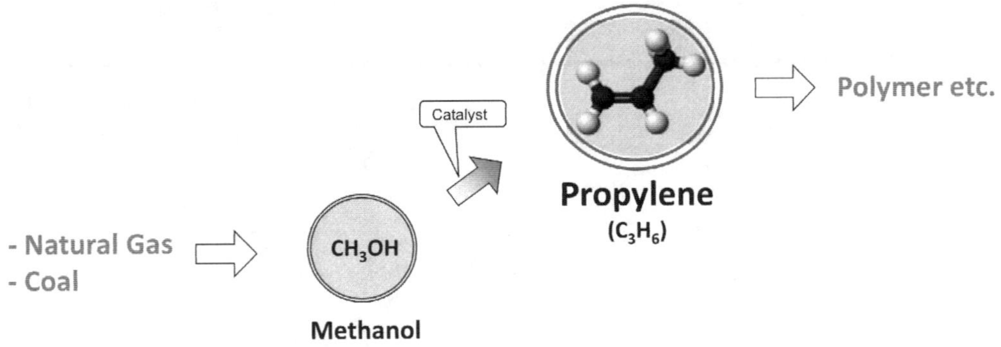

-On-Purpose Propylene Production from Methanol

JGC CORPORATION | - 2 - | MITSUBISHI CHEMICAL

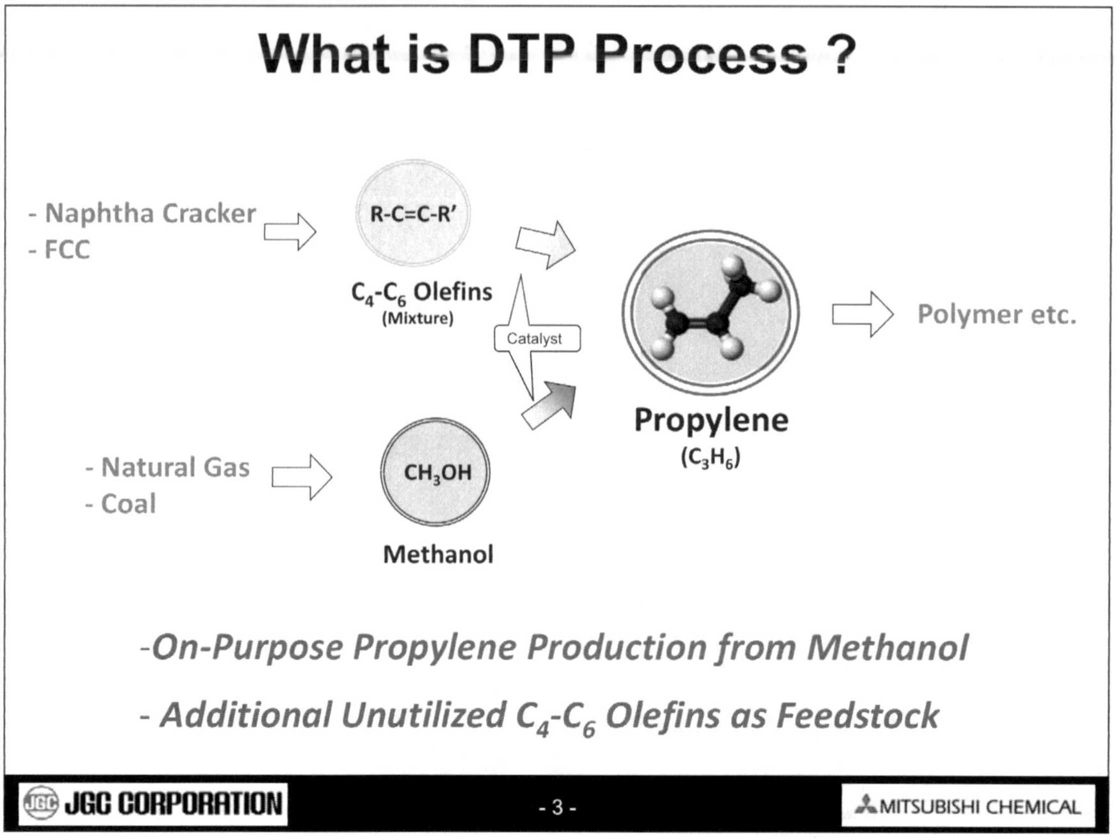
What is DTP Process ?
- Naphtha Cracker
- FCC
R-C=C-R'
C_4-C_6 Olefins
(Mixture)
Catalyst
Propylene
(C_3H_6)
Polymer etc.
- Natural Gas
- Coal
CH_3OH
Methanol
-On-Purpose Propylene Production from Methanol
- Additional Unutilized C_4-C_6 Olefins as Feedstock
JGC CORPORATION
- 3 -
MITSUBISHI CHEMICAL

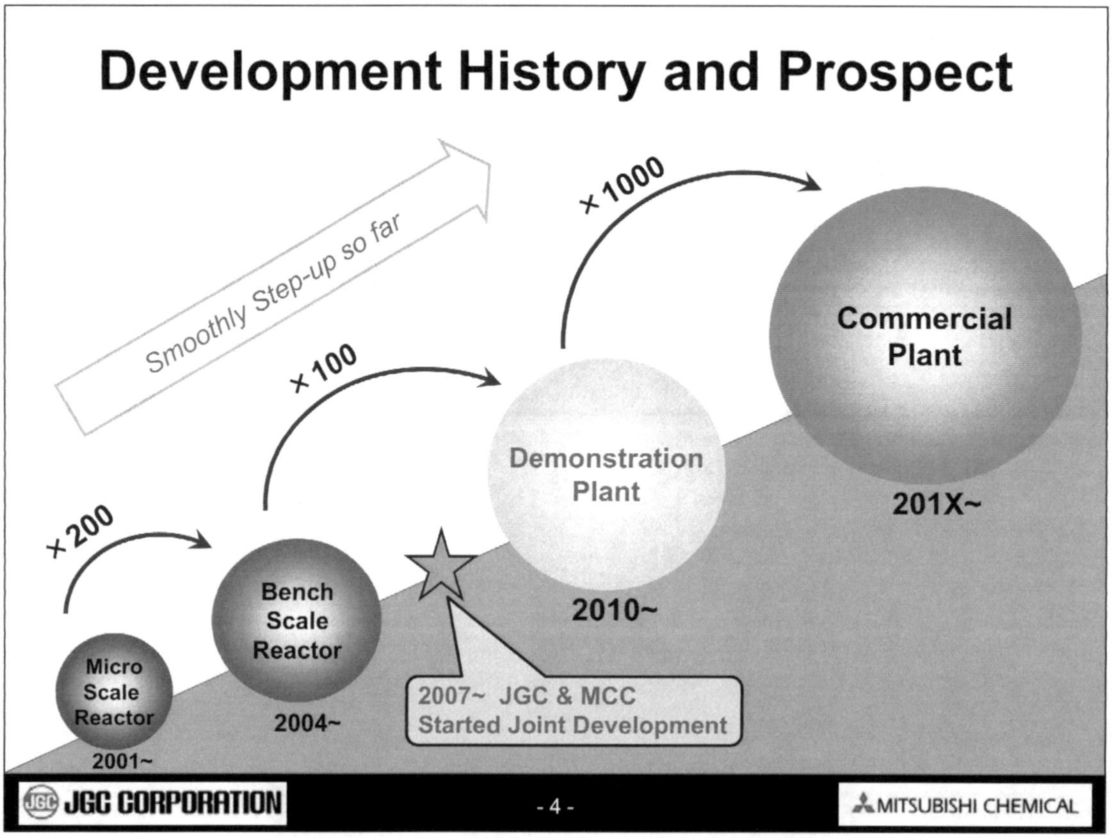
Development History and Prospect
Smoothly Step-up so far
×200
×100
×1000
Micro Scale Reactor
2001~
Bench Scale Reactor
2004~
Demonstration Plant
2010~
Commercial Plant
201X~
2007~ JGC & MCC
Started Joint Development
JGC CORPORATION
- 4 -
MITSUBISHI CHEMICAL

Demonstration Plant Overview

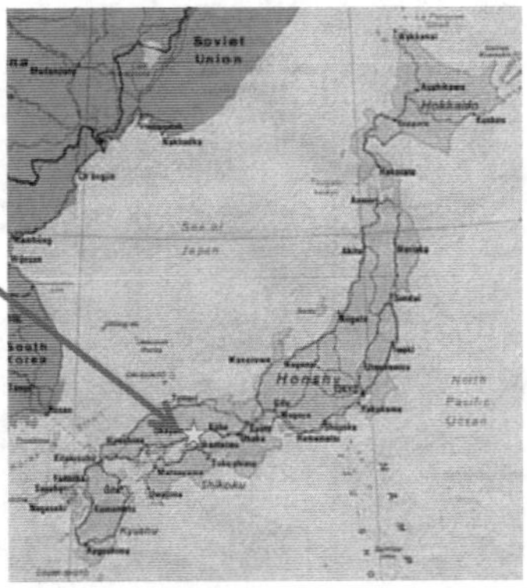

- Located in Mitsubishi Chemical Mizushima Plant
- Successfully Start-up in August 2010
- Continuous Operation for Two Years

Process Flow (Standalone Case)

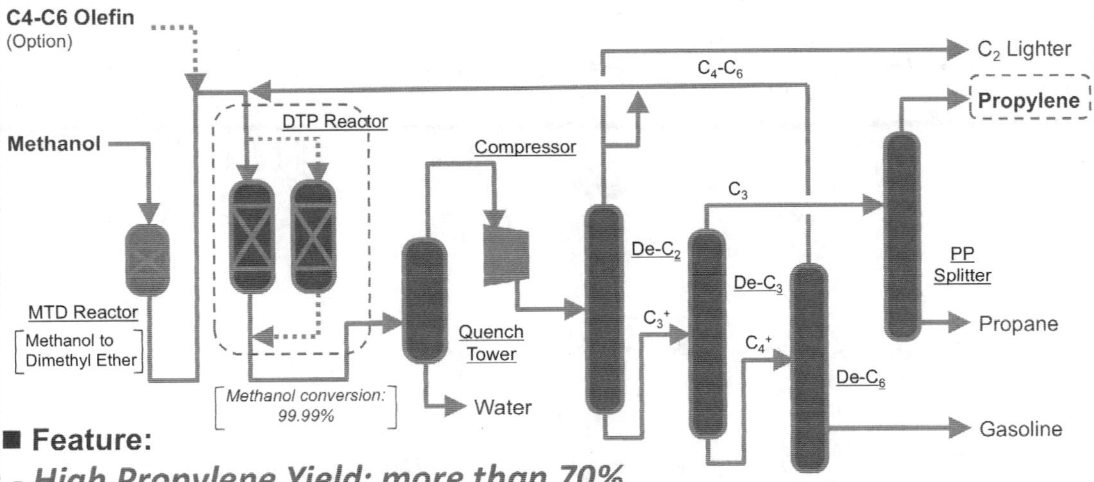

■ **Feature:**

- ***High Propylene Yield: more than 70%***
- ***Applied Existing Proven Technologies***

Process Flow (Integration Case)

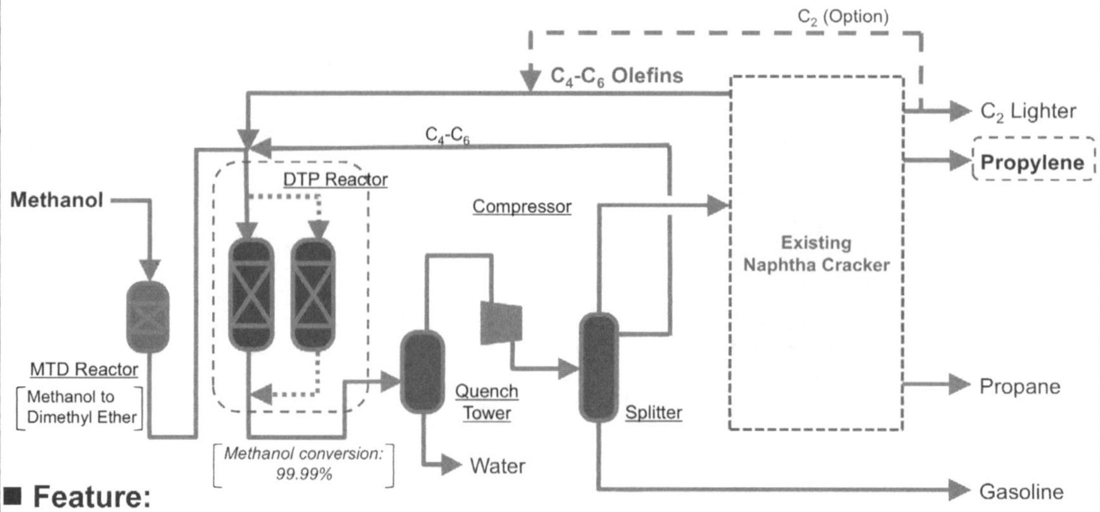

■ **Feature:**

- ***High Propylene Yield: more than 70%***
- ***Applied Existing Proven Technologies***
- ***Low Investment Cost in case of Integration with Naphtha Cracker***

Application

(1) Naphtha Cracker Integration Case (Propylene Capacity: 20KTA – 200KTA)

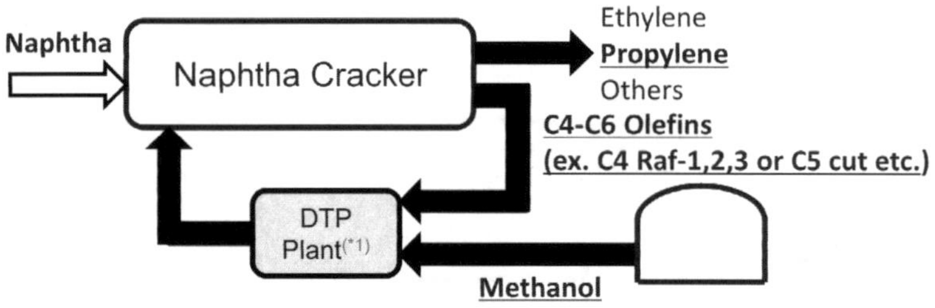

(*1) Utilizing facilities in Naphtha Cracker.

(2) Standalone Case (Propylene Capacity: 250KTA – 500KTA)

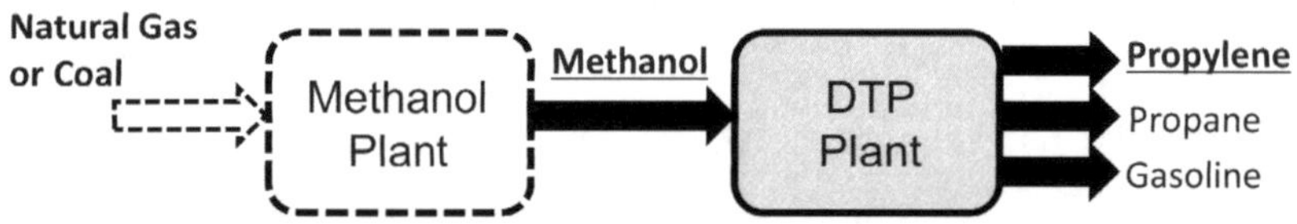

- ***Two Cases available depending on Client requirements !!***

DTP is...

- *High Yield Propylene Production Process*
- *Flexible Feed Methanol to Olefin Process*
- *Process that increase P/E Ratio for Cracker*

Thank you very much !!

Contact:
Mitsuo Morita
Market Development Dept.
Global Marketing Division
2-3-1, Minato Mirai, Nishi-ku,
Yokohama 220-6001, Japan

Phone: 81-45-682-8371
FAX: 81-45-682-8183
E-Mail: morita.mitsuo@jgc.co.jp

352533 Self-Cleaning "Bernoulli" Type Filters Used in Onshore and Offshore Applications

Abstract:

Georg Schuenemann GmbH (SAB-BREMEN) has developed a new Filter Type F-480, dubbed the "Compact Giant" for applications in onshore water intake and cooling water systems, sea water desalination plants, and in offshore projects, such as FPSO's, Drillships, and Production Platforms.

This paper will describe the essential design features of the filters, based on the "Bernoulli principle" which result in a fully automated self-cleaning filter system, and choice of materials in a sea water service with high corrosion rates.

A case study will be presented in which stringent project requirements such as substantial flow rates, adaptibility to space limitations and pipe connections had to be observed. Reliability of the system, extended run times without the need for frequent replacement of filter elements, as well as ease of monitoring and operator involvement have been achieved.

352653 Effects of Support on Sulfur Tolerance and Regeneration of Pt Catalysts Measured By Ethylene Hydrogenation and EXAFS

Abstract:

We investigated the effect of support on sulfur tolerance and regenerability under reducing environments by measuring the rate of ethylene hydrogenation, combined with hydrogen chemisorption, and extended x-ray absorption fine structure (EXAFS). Alumina and titania (P25) supports were used to prepare Pt catalysts at 1 wt.%. The catalysts were tested after treatment in H_2S/H_2 at 250 °C followed by regeneration treatments in H_2 at 250 °C, 350 °C and 450 °C. The sulfided catalysts exhibited a 20-27 times decrease in the rate of ethylene hydrogenation and a 4-6 times drop in the Pt surface area. Regenerations up to 450 °C were unable to remove all the sulfur, as shown by the presence of Pt-S bonds by EXAFS. However, a partial recovery of the hydrogenation rate per mole of Pt was observed on sulfided Pt/Al_2O_3 after reduction at 450 °C, while the induction of strong metal support interactions (SMSI) at reduction temperature above 350 °C was observed on Pt/P25, regardless of the presence of sulfur. For Pt/P25, the reversal of the SMSI state together with sulfur removal by mild oxidation suggests that sequential reduction/oxidation treatments may be more effective in restoring the S-free state of TiO_2-supported catalysts.

352799 Improved Operational Efficiency and Reliability through Insulation Materials Selection

Abstract:

This paper will explore the implications of insulation materials selection on the energy efficiency and reliability for Ethylene production units. Considerations will include insulation properties, structural support, installation methods, installation and maintenance costs.

BOILER MACT COMPLIANCE: MERCURY IN FUEL GAS

PATRICK LAINE, WES YONGE, & VIC VICKERY
PEI, HOUSTON, TX 77043

Patrick Laine, Ph.D.
Chief Scientist
Houston, TX
850-866-6686
plaine@pei-tx.com

Wes Yonge, PE
Chief Engineer
Houston, TX
713-806-8396
wyonge@pei-tx.com

James 'Vic' Vickery
Chief Technology Officer
Houston, TX
281-536-0899
vvickery@pei-tx.com

1 Introduction & Objectives

For most ethylene producers the primary objective of this sampling and analysis effort will be to verify and "establish a mercury fuel specification through which gas-fired boilers that use a fuel other than natural gas or refinery gas may be considered Gas 1 unit" as identified in 40 CFR Part 63.7521.(f).

Measurement and Monitoring Requirements: Decision Rules

❑If Hg concentration ≤ 50% of Gas 1 specification* then no further testing required

❑If Hg concentration > 50% and ≤ 75% of the specification* then semi-annual sampling required

❑If Hg concentration > 75% of the specification* then monthly sampling required

*Specification is **40 μg/Sm³**

The goal will be to execute a sampling and analysis campaign that will produce compliance quality data. The key to producing and validating compliance quality data is a well-designed and executed QA/QC protocol. At a minimum, a performance based QA/QC protocol is recommended that includes the following components:

- Sorbent Trap Breakthrough Bed:
  - ❑ Sorbent traps consisting of a minimum of two sorbent beds in series so that the trailing bed can be analyzed to verify that no breakthrough occurred from the leading bed.
- Duplicate Samples:
  - ❑ The sampling system should be configured with two trains to accommodate the collection of simultaneous duplicate sorbent trap samples, in parallel, providing a means of verifying precision in the sample collection process.
- NIST (National Institute of Standards and Technology) Traceable Spike:
  - ❑ Sorbent traps with a third bed (downstream) that is spiked with a NIST traceable spiked mass of mercury serves as a matrix spike and provides a means of verifying that components in the matrix (the fuel gas) are not affecting the trapping efficiency of the sorbent trap.

2 Methods

Sorbent Trap Mercury Sampling Methods Comparison

Method	Description	Method Type: S = Sampling A = Analysis	Evaluation	Sample Collection Phase QA/QC
ISO 6978	Hg in Natural Gas by Thermal Desorption (CVAFS)	S & A	Method calls for low sample volumes can lead to non representative results.	No
EPA Method 30B	Determination of Gaseous Hg Emissions from Coal Fired Combustion Sources using Carbon Sorbent Traps	S	Requires modification (Modified M30B)	Yes
ASTM D6350	Hg S&A in Natural Gas (AFS)	S & A	Method calls for short term samples (i.e., low sample volume) which leads to poor accuracy.	No
ASTM D5954	Hg S&A in Natural Gas (AAS)	S & A	AAS is less sensitive than AFS.	No
EPA Method 1631	Measurement of Hg in Water	A	No	Yes

3 EPA Method 30B QA/QC Performance Criteria

QA/QC Test or Specification	Acceptance Criteria	Frequency
Analyzer calibration	Correlation Coefficient: 0.999	Daily
Transfer efficiency of analyzer	%Transfer Efficiency: 90%–110%	Once per sorbent trap
Initial calibration blank	< 0.5 ng	Once before each calibration
Sorbent Trap Section B breakthrough	≤ 10% of Section A Hg mass, for Hg concentrations > 1μg/dscm ≤ 20% of Section A Hg mass, for Hg concentrations ≤ 1μg/dscm	At each sample point
Spike Recovery Test	Average recovery between 75% and 125 % for Hg_0	Once per sample point
Paired Sorbent Traps Agreement	≤ 10% RD for Hg concentration > 1μg/dscm ≤ 20% RD or 0.2 μg/dscm absolute difference for Hg concentrations ≤ 1μg/dscm	At each sample point

4 Equipment

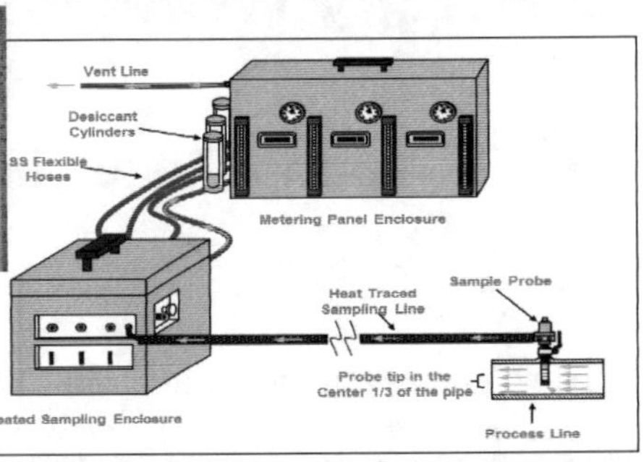

Flow

	Primary Bed	Breakthrough Bed	Spiked Bed

QA/QC Acceptance Criteria:	99% of Total Collected Hg	≤ 10% of Primary Mass	Recovery between 75% and 125%

Gold Trap:

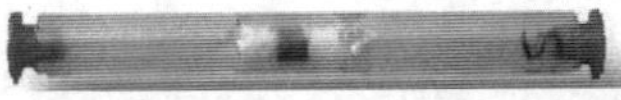

5 Conclusions/PEI Recomendations

❑ There are five methods accepted by the Boiler MACT rule. PEI has extensive experience and expertise employing **all five** of these methods.

❑ Operators can achieve compliance without submitting a Fuel Analysis Plan by using ASTM D6350, however, we have noticed variable results with this method.

❑ The **PEI recommendation** for obtaining the most accurate compliance quality data is to use Modified EPA Method 30B in conjunction with EPA 1631 where traps should be analyzed in a certified ISO 17025 laboratory. The recommended pathway includes a rigorous, performance based QA/QC protocol that validates accuracy and precision of all sample collection procedures.

351376 Young Professional Simulation Tutorial

Abstract:

18C00 Young Professional Simulation Tutorial

Authors: Naomi Hua and Mike Donahue, Invensys

Process simulation is a complex skill to master. Even with years of experience, a typical user is constantly facing new problems that need to be addressed. There are many issues that can lead to unexpected and unnecessary errors with the use of process simulation tools. Regardless of what simulation program you are using, these issues are common from one program to another. In this presentation, targeted for the young professional, we will discuss the "Most Common Pitfalls in Process Simulation". For each topic, we will discuss various techniques and methods these troublesome problems and prevent their occurrence.

Topics to be discussed include:

1. Erroneous data entry
2. Simulation defaults
3. Bad plant data
4. Thermodynamic method selection

5. Simulation tolerances and convergence

6. Recycle identification and handling

7. Model complexity

8. Difficult or impossible specifications

9. Assay pseudo-component development

10. Assay property blending

1. Erroneous data entry is the most common mistake made when building a process simulation model. This presentation will cover methods that can help ensure proper data entry in your process simulation model, such as checking the data you entered, particularly the units of measure, using "alternate forms" of data input such as copy/paste and data import, and reviewing all warning and error messages.

2. Defaults are methods, values, and conditions that are automatically selected for you by the program when you begin a new simulation model. These defaults will affect the results calculated by your model and every simulation program has them. Most defaults within simulation programs fall under categories such as convergence criteria, thermodynamic methods, tolerance and units of measure. The problem with defaults is unintended results from not fully understanding the effects a particular default has on the model. Take the time to understand the defaults of the program and how these defaults will affect the results. If you do not agree with the defaults selected by the program, most simulation programs will allow the user to change the default.

3. Plant data changes over time. The feed information and process conditions of the unit operations may be different. Not only does plant data change over time, but measurement information is often very inaccurate. This could be due to faulty or poorly calibrated measurement equipment, leaks in lines, sensor failures, and other issues. Because of these changes and inaccuracies over time, you always have to validate the information that is entered in your model to ensure that you are obtaining the proper results. To do so, you can utilize programs that can help with data validation and reconciliation like SimSci's ROMeo online optimization program. With it, you can connect your model to your data historian to constantly import current data measurements into your model and then reconcile these measurements to determine when problems arise within your plant and where those problems are occurring. You will also need to use your engineering judgment to determine if the data is valid. More details will be provided during the presentation.

4. The proper selection of thermodynamic methods for your simulation is the most important decision that you make when modeling a process. The method you select could have a drastic effect upon your results. Often, any number of thermodynamic methods is suitable for a given application. It is up to the user to determine the best method to use based upon knowledge, personal experience, and best judgment. Some important factors that need to be considered are the components being used and the operating conditions.

Oftentimes, the selected thermodynamic method still needs to have additional inputs such as binary interaction parameters to simulate the process accurately.

5. Many of the calculations in process simulation are iterative. When these calculations iterate, the program compares the results calculated from one iteration to the results calculated in the next iteration in order to determine

whether it is converged to a solution or needs to continue calculating. This method of comparing results is where tolerances come into play and simulation programs use tolerances to determine when a solution is reached. One of the major mistakes when modeling a process is not understanding or using the correct tolerance values. A looser tolerance may make the file easier and quicker to converge, but it also reduces the accuracy of the model. A tighter tolerance may be more accurate, but might make it impossible for the model to solve or increase the convergence time exponentially. A common practice is that when you have reached a converged solution with your model, tighten the tolerances within the model and solve it again. If you notice that your answers have changed significantly, you need to continue tightening the tolerance values until your results change little from one iteration to another. You will then know that you have reached an accurate solution for your model.

6. Recycles in process simulation occur when you return to a previously calculated unit operation to recalculate the unit due to a change made elsewhere in the model. They are very common and can often be difficult to converge. Some suggestions on troubleshooting recycle convergence include initially solving your model without the recycles to see how it behaves, providing initial estimates for recycle streams, using tighter tolerances on unit operations within recycle loops than the total recycle tolerances themselves, and ensuring that all components have a way in and out of a recycle by performing mass balances and using purge or makeup streams where necessary.

7. The general rule of thumb when modeling is to start simple and add complexity as you go. Each time you add a few streams or unit operations converge and save the model before moving on. Make sure you justify any additions that you make to the model. You can most likely model your system without including every unit from your PFD in the simulation. When modeling, think about the goal of your simulation and not about literally translating every real-world piece of equipment into the simulation model. Make the judgment call on what is most important and what the purpose is of every unit you add to your simulation. For example, if your real-life process shows a heat exchanger, a pump and a splitter, but all that is necessary is that the temperature, pressure, and separation are achieved properly, consider using a flash drum to accomplish all three.

8. Specifications within process simulation are everywhere and every stream and unit operation you place in your model contains specifications. In its simplest form, a specification is a constraint that you place on the model that the model needs to meet in order to successfully converge. For streams, these are usually temperatures, pressures, flowrates and compositions. For unit operations, they range from process conditions, like temperatures and pressures, to product conditions, like an overhead composition in a column. There are some pitfalls, however, that you can run into with specifications such as azeotropes in columns, high purity specifications, and the possibility of multiple solutions.

9. Petroleum assays are sets of distillation data that are used to characterize all of the components that reside in a given crude sample. When you enter distillation data into a process simulator, the program converts it into components that it can use when modeling your process. The components that are generated are called pseudo-components, and each pseudo-component represents all of the unknown components that would be within a given boiling point range in the crude. Some of the things you will want to ensure when properly developing your assay components are did you use the correct distillation curve basis, did you enter the data correctly, did you use enough cuts and are you using the correct characterization methods .

10. When you have more than 1 assay in a simulation model, you need to be concerned with assay blending and how it will affect the results of your file. Assay blending is the mixing of 2 or more groups of components (i.e. ¨C sets of cuts/pseudo-components) in order to calculate averaged component properties for cuts that are in the same boiling point range as one another. Once these new averaged component properties are determined, the program will use these averages from this point on in the simulation model any time it needs to access component

properties for a given pseudo-component. All component properties are blended, including normal boiling points, specific gravity, molecular weight and refinery inspection properties, and usually the blended component properties are calculated using a weighted average

These topics are based on experience and feedback from end-users to technical support representatives for process simulators. They often represent the most commonly asked questions or areas that most users find to be problematic. While the specific examples highlighted in each topic come from SimSci software, the intent of this tutorial presentation is to be broad-based and is particularly applicable to young professionals seeking to develop their experience with simulation.

Polymeric Sulfonic Acid Composites for Fuel Cell Applications: Effects of Casting Solvent and Sonication Time on Membrane Performance

*Jimoh K. Adewole[1], Abdullah S. Sultan[1, *], Amir Al-Ahmed[2], and S. M. Javid Zaidi[3]*

[1]Center for Petroleum & Minerals, King Fahd University of Petroleum & Minerals, Dhahran 31261, Saudi Arabia

[2]Center of Research Excellence in Renewable Energy, King Fahd University of Petroleum & Minerals, Dhahran 31261, Saudi Arabia

[3]School of Chemical Engineering, The University of Queensland, Australia

Abstract

Polymer blends comprising sulfonated poly (ether ether ketone) (SPEEK) and 3wt% (2-acrylamido-2-methyl-1-propanesulfonic acid) (PAMPS) were prepared by solution casting method. Two solvents (N,N-Dimethylacetamide (DMAc) and N-Methyl-2-pyrrolidone (NMP)) were uesd separately. Residual solvents in the membrane samples after drying were quantified using thermo gravimetric analysis. Effects of periods of sonication (0 to 90 minutes) and types of solvent on the membrane performance were evaluated interms of morphology and water uptake. An attempt was made to correlate membrane performance with the time of sonication in order to select the most suitable solvent and membrane casting conditions for subsequent membrane preparations. For both solvents, the water uptake was observed to first decrease at 10 minutes of sonication and then increased as the sonication time was increased. The minimum water uptake recorded for DMAc samples was 19.23% at 20 minutes sonication time while that of NMP sample was 24.24% at 10 minutes. However membrane morphology was found to deteriorate with increase in sonication time. No correlation was obtained between the membrane properties and the sonication time. In general, membrane performance was better for membranes samples using DMAc than NMP solvent. Therefore, DMAc was selected as the casting solvent for membrane preparation on which further studies will be carried out.

Keywords: morphology, water uptake, residual solvent, thermogravimetric analysis

Introduction

Water soluble polymers have received academic and industrial interests due to their potential applications in fuel cells and other electrochemical devices (Ye et al. 2012). One of the water soluble polymers that have attracted publication interest is poly (2-acrylamido-2-methyl-1-propanesulfonic acid) (PAMPS) (Florjańczyk et al. 2003). Generally, some of the major performance criteria for fuel cell materials include proton conductivity, water and methanol uptake, good mechanical properties, chemical stability, thermal stability and methanol permeability (Gil et al. 2004; Swier et al. 2005; Ye et al. 2012). PAMPS was reported to posses higher proton conductivity than partially hydrated Nafion. In addition, it has been found to be more tolerant to water fluctuation in high temperature operations. Some of the major drawbacks in the use of PAMPS are its poor mechanical properties, thermal stability, water and methanol uptake and solubility in many solvents (Florjańczyk et al. 2003). One of the approaches that have been proposed to solve these problems is development of polymer composites.

This research paper investigates the synergistic effect of combining water soluble PAMPS with sulfonated poly (ether ether ketone) (SPEEK) which is non - water soluble polymer to prepare composite membranes for application in fuel cells. Specifically, preliminary results on the performnance of the composite were presented. Performance were evaluated based on the composite morphology and water uptake. Effects of sonication time and types of casting solvent on membrane properties were also evaluated.

Material and Method

Material

Sulfonated poly (ether ether ketone) (SPEEK) purcahsed from FuMA-Tech, Germany, and poly (2-acrylamido-2-methyl-1-propanesulfonic acid) (PAMPS) purchased from Aldrich Chemical Co., are the two major polymers used in this research. The SPEEK has a DS of 0.533 and solubility of 99% in DMAc. SPEEK is derived by sulfonation of Poly (ether ether ketone) (PEEK). PEEK is a commercially successful polymer due to its excellent thermal stability and solvent resistance (Mohr et al. 1991). For sometimes, further application of PEEK have been stalled due this solubility issues. By sulfonation, PEEK can be made soluble in most commercial solvents. Importantly, the solubility of SPEEK also depends on its degree of sulfonation (DS). Since, this work involves the use of two polymers, a solvent which will disolve both the PAMPS and SPEEK must be selected. Literature review showed that PAMPS dissolves in N,N-Dimethylformamide (DMF), N,N-Dimethylacetamide (DMAc), propylene carbonate (PC), and water (Florjańczyk et al. 2003). Initial solubility studies that was carried out in our laboratory was done for N,N-Dimethylacetamide (DMAc) and N-Methyl-2-pyrrolidone (NMP). The choice of NMP was due its unique solvent properties and its structtural distiction from DMF and DMAc (Figure 1). Moreover, no literature was found where NMP was used as casting solvent for PAMPS. Thus, DMAc and NMP were used as solvents in this work. All materials were used as received from the manufacturers.

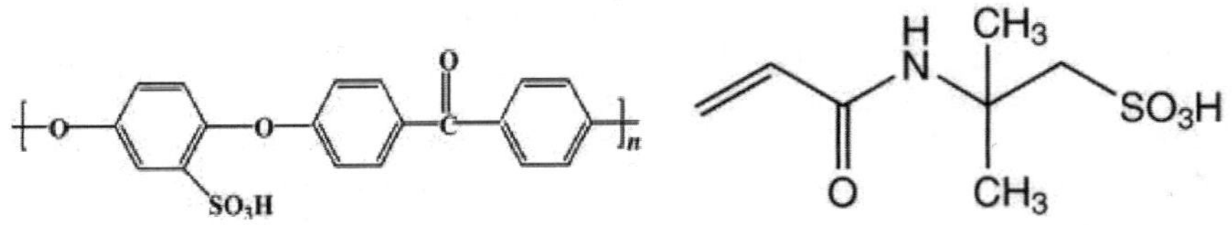

Figure 1 Structure of SPEEK and PAMPS

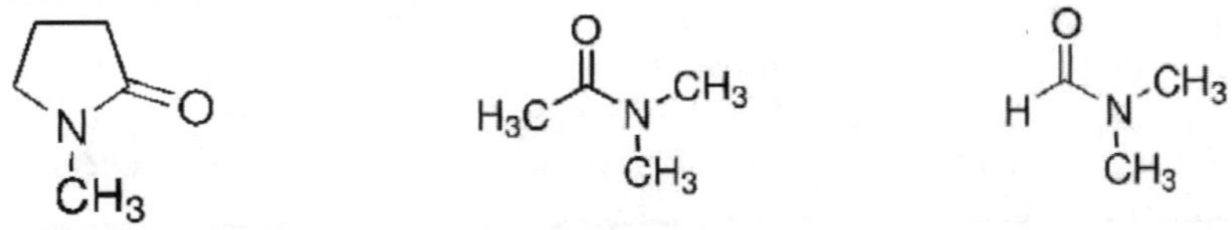

N-Methyl-2-pyrrolidone N,N-Dimethylacetamide N,N-Dimethylformamide

Figure 2 Structure of Solvents

Experimental Procedure

Membrane Preparation

SPEEK was first dissolved in solvents (DMAc and NMP) and stirred for 2hours at 70°C. PAMPS was then added and the blend was stirred for 24hours. The solution was sonicated for various periods of time ranging from 10 to 90minutes using an Ultrasonic bath. In untralsonic bath, the final temperature achieved for a given time of sonication depends on the tempearutre of the laboratory (summer or winter). It is therefoe genrally advisable that maximum temperature of the bath is determined and maintains (as the so - called equilibrium tempearture), when operating continuously under ambient conditions (Santos et al. 2009). During the sonication, equilibrium temperature was maitained by simply filling the bath with water that was preheated to sample temperature.

Membrane Characterization

Water uptake and morphological tests were performed. Water uptake was carried out by first vacuum drying all membrane samples at 100°C until constant weights were obtained. Dried samples were then immersed in deionized water for 24h at room temperature (25°C). Samples were taken out of the water after 24h, wipped to remove surface water and then weighed immediately. Water uptake was calculated using Eq. 1.

$$water - uptake = \frac{W_{wet} - W_{dry}}{W_{dry}} \qquad \text{Eq. 1}$$

Surface morphology of the membrane was done using scanning electron microscopy (SEM). Thermogravimetric analysis (TGA) was done from room temperature to 350°C using TA Q600 analyzer at heating rate of 20°C/min under nitrogen atmosphere with flow rate of 20mL/min using ~ 8 -10mg membranes samples.

Results and Discussion

Membrane Morphology by Scanning Electron Microspcopy (SEM)

Results of surface morphological analysis performed on membrane were shown in Figure 3 and 4. The overall pictures of this figure depicts a dense membrane with some surface defects observed for most of the sonicated samples. For example, Figure 3A showed some kind of layered arrangement with no visible pores at the surface. Also, the PAMPS seems to be well distributed throughout the SPEEK. The micrograph in Figure 3B can be observed to be rough and possesses some uneven ramdom distribution of PAMPS agglomerates indicating that the distribution of PAMPS have been disturbed by the 10minutes sonication. An increase in sonication time to 20 minutes was observed to improve the distribution of PAMPS. This was however accompanied by some cracks believed to be due to the slight increase in the viscousity of the blend solution. The vicousity increase was brought about by the excape of some fraction of the solvent during sonication. As the sonication time was increased, PAMPS was observed to become more evenly distributed and the cracks become more visible and larger in size (Figure 3D to F). SEM images of samples at sonication time of 10 and 30 minutes were reproduced in higher magnification of 10μm (Figure 4) in order to further revealed the effect of sonication on the morphology of the somposites. Similar morphological behaviour has been reported in literature, though for inorganic/organic clusters of nanoclay/epoxy composites membranes (Lam et al. 2005).

Water Uptake

Water uptake have been reported to be affected by the degree of sulfonation, membrane pretreatment, hydration, ambient relative humidity and water temperature (Xing et al. 2004; Wu et al. 2006). Figure 5 showed the effect of sonication time and solvent type on water uptake that was measured at room temperature (25°C). For both solvents, water uptake is same at zero sonication time. There was however a further decrease in the water uptake for NMP at 20 minutes sonication time before it beging to increase. Thus, the best sonication time for both solvent lied between 10 and 20 minutes with respect to water uptake. Interestingly, this result correspond to that reported by Lam (2005) on the effect of ultrasonication on microhardness of composites.

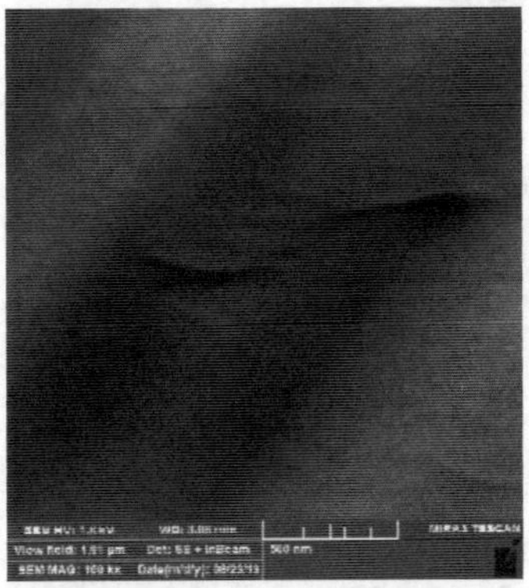

A) No sonication

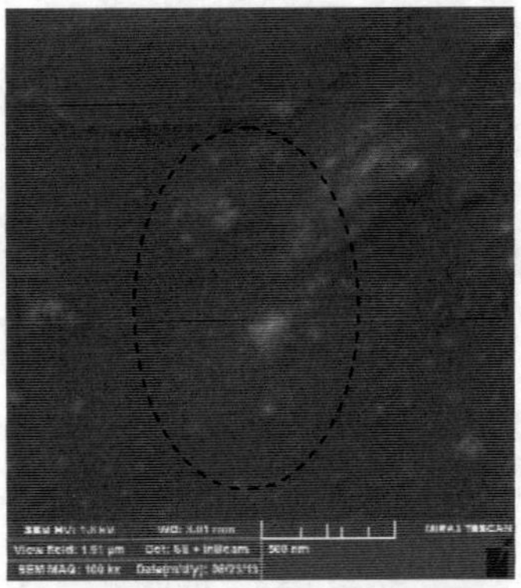

B) 10 minutes sonication

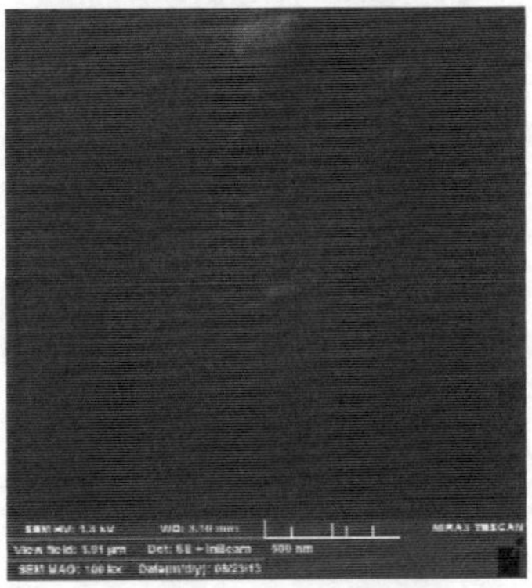

C) 20 minutes sonication

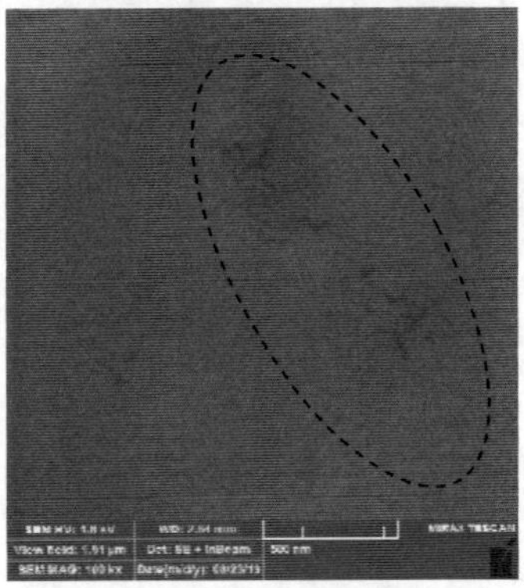

D) 30 minutes sonication

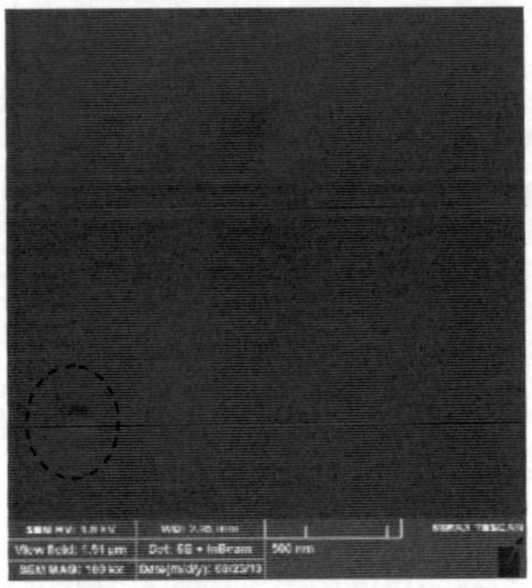

E) 60 minutes sonication

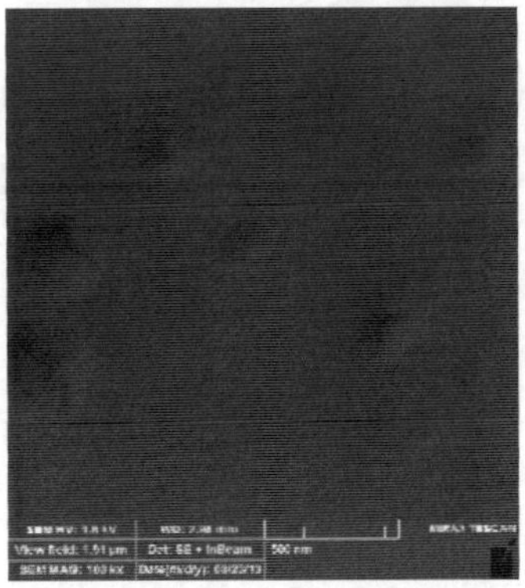

F) 90 minutes sonication

Figure 3 SEM micrograph of SPEEK-PAMS composite with various times of sonication with magnification of 500nm

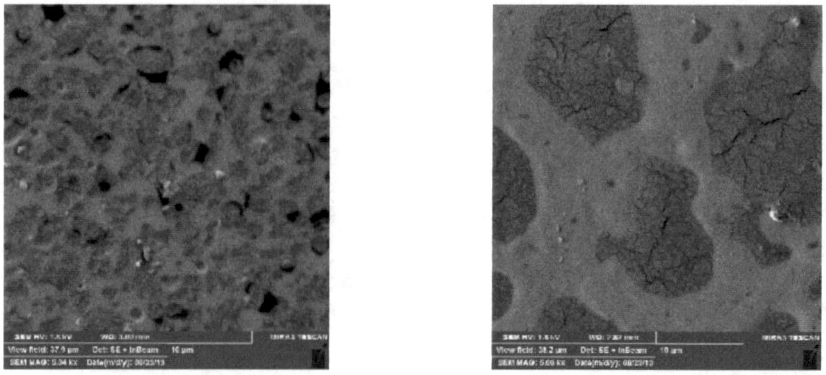

A) 10 minutes sonication B) 30 minutes sonication

Figure 4 SEM micrograph of SPEEK-PAMS composite with various times of sonication with magnification of 10μm

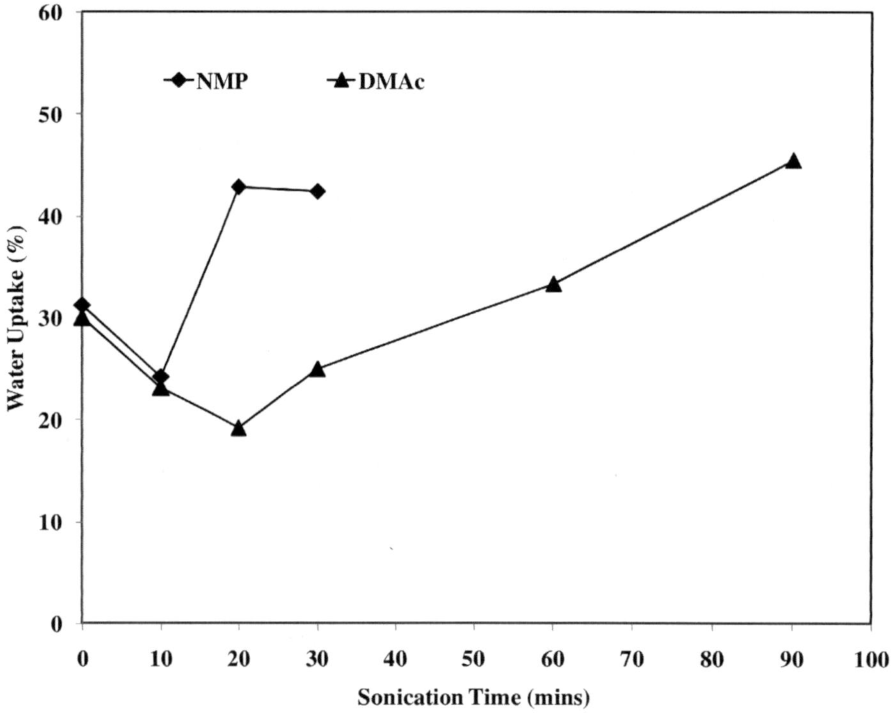

Figure 5 Water uptake of SPEEK-PAMS (3wt%) composite with variuos sonication times

Thermogravimetric analysis (TGA)

The difference in membrane performance resulting from the use different casting solvents have been attributed to the presence of residual solvents in the membrane samples after evaporation (Joly et al.

1999). TGA was thus used to quantify the residual solvent and evaluate its effect on the membrane performance as has been previously demonstrated in lieterature (Recio et al. 2007; Iqbal et al. 2008). Figure 6 shows the TGA curve for samples prepared from NMP and DMAc with different post vacuum heat treatment conditions. From this figure, weight loss can be attributed to evaporation of variety of substances, including solvents, from the samples (Rich et al. 1997). Table 2 contains the solvent and its boiling points, and weigth loss at the boiling point range that were extracted from TGA results. For sample heat treated for 24hrs, the overall percentage weight loss up to 204°C was about 8wt% of membrane sample for NMP while it is about 6.3wt% for DMAc. These will definitely include compnenents such as water, solvents and other unidentified substances. To quantify residual solvents, weight loss was estimated between 192 -214°C for SPEEK-PAMPS-NMP and between 154 and 176 for SPEEK-PAMPS-DMAc. The results obtained showed that the amount of the residual solvent in NMP is larger than that of DMAc by 1.25wt% (Table 1). To further investigate if this difference is responsible for higher value of water uptake for the NMP sample, membrane samples made from both solvents were further heat treated under vacuum for 72hrs. Residual solvent after this treatment was slightly higher in DMAc than NMP. Despite this, the value of water uptake of the NMP membrane is still higher than that of DMAc by approximately the same value as before the heat treatment (1.71wt%).

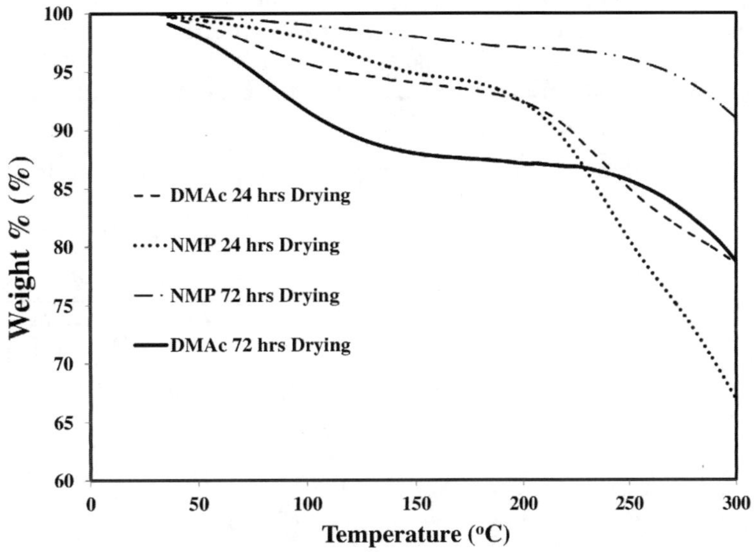

Figure 6 TGA of SPEEK-PAMPS-NMP and SPEEK-PAMPS-DMAc

Table 1 Residual solvent in SPEEK-PAMPS composites membrane samples

Membrane Sample	Solvent	B. Pt. (°C)*	Residual Solvent (wt%)	Water Uptake
DMAc 24 hrs Drying	DMAc	164.5 – 166	0.57	30.00
NMP 24 hrs Drying	NMP	202 – 204	2.85	31.25
DMAc 72 hrs Drying	DMAc	164.5 – 166	0.367	21.21
NMP 72 hrs Drying	NMP	202 – 204	0.217	22.92

*Boiling Points of Solvents

On a general note, results from Figire 1 – 5 showed that the best perfomance can be obtained for variety of sonication time depending the property of preference. Also, performance of DMAc solvent was observed to be better than that of NMP on the basis of water uptake. For this reason, DMAc was selected as the casting solvent for membrane preparation on which further studies will be carried out. Water uptake of DMAc membrane, prepared without sonication, was comapred with those reported in previuos literature. Table 2 illustrates the comparison between SPEEK-PAMPS and other relevant membranes previuosly reported in literature. In addition, this table serve to provide insight into the possible potentials of the composites. From this table, it was observed that the water uptake for SPEEK-PAMPS is within the range of values of other membranes on the table. Its water uptake is less than that of Nafion 112 which is a commercial membrane. The differences observed between the water uptake values of SPEEK – PAMPS and other membranes from the table can be explained based on three important factors - SO_3H group, the degree of sulfonation and experimental temperature (Gil et al. 2004; Shan et al. 2006; Wu et al. 2006). Water uptake of SPEEK – PAMPS is higher due to the presence of SO_3H group in the PAMPS molecules (Figure 2). SO_3H group have been reported to enhance membrane acidity and hydrophilic properties (Shan et al. 2006). Infact, the higher the SO_3H group the larger the water uptake. The degree of sulfonation has been reported to affect the water uptake of membranes. The higher the degree of sulfonation, the larger the value of water uptake (Gil et al. 2004). Thus, water uptake of SPEEK – PAMPS is lower than SPEEK -3 due their different degree of sulfonation, 53.3 and 72% respectively and the experimental temperature 24 and 60°C, respectively. The experimental temperature of SPEEK – 2 is close to that of the composite. However, the results cannot be compared in this case since the DS of SPEEK – 2 was not known. Above explanation is also true for the values of water uptake for SPEEK – 1.

Table 2 Comparison of water uptake values obtained with those from Literature (Xing et al. 2005)

Membrane	Water Uptake (wt%)	Experiment Temperature (°C)	**D S (%)	Reference
Nafion 112	22.85	N/A	-	(Xing et al. 2005)
SPEEK – 1*	16.00	30	45	(Ren et al. 2005)
SPEEK – 2*	44.90	25	N/A	(Zhong et al. 2008)
SPEEK – 3*	85.00	60	72	(Wu et al. 2006)
SPEEK– 4*	15.00	N/A	N/A	(Handayani and Dewi 2011)
SPEEK/PAMPS (3wt%)	21.21			This work

*The numbering have been included for explanation purposes.
**DS means Degree of Sulfonation

Conclusion and Recommendations

Blend of SPEEK and PAMPS was prepared by solvent casting method and its performance was evaluated with respect to its morphology and water uptake. Effect of sonication time (10 – 90minutes) and solvent types (DMAc and NMP) were also investigated. Performance of membrane prepared using DMAc was observed to be better than that of NMP. There was no correlation between the sonication time and membrane properties, but it was observed that the sonication deteriorate the membrane morphology. Membrane prepared without sonication shows better surface morphology, whereas the best water uptake was observed between 10 and 20 minutes. Thus, other properties need to be investigated in order decide the best sonication time. Therefore, further investigation such as thermal stability, methanol permeability, ion exchange capacity and gas (O_2 and H_2) permeability of the membrane are recommended for a complete evaluation of the membrane performance for direct methanol fuel cell (DMFC) application. All these investigations are in progress in our laboratory.

Acknowledgement

Authors would like to thank the National Science, Technology and Innovation Program (NSTIP) of KACST for funding this research under project number 10-ENE1374-04. Also, authors extend their appreciation to Center of Petroleum & Minearls, Center of Excellence in Renewable Energy, Mr. Dan Jerry D.Cortes of the Nanotechnology Laboratory, Chemical Engineering Department, King Fahd University of Petroeleum & Minerals, Saudi Arabis for their supports.

REFERENCE

Florjańczyk, Z., et al. (2003). "Proton conducting electrolytes based on poly(2-acrylamido-2-methyl-1-propanesulfonic acid)." Electrochimica Acta **48**(14–16): 2201-2206.

Gil, M., et al. (2004). "Direct synthesis of sulfonated aromatic poly(ether ether ketone) proton exchange membranes for fuel cell applications." Journal of Membrane Science **234**(1–2): 75-81.

Handayani, S. and E. L. Dewi (2011). "Influence of Silica/Sulfonated Polyether-Ether Ketone as Polymer Electrolyte Membrane for Hydrogen Fueled Proton Exchange Membrane Fuel Cells." Internat. J. of Sci. and Eng. **2**(2): 26-30.

Iqbal, M., et al. (2008). "Solvent effect on morphology and CO2/CH4 separation performance of asymmetric polycarbonate membranes." J. Membr. Sci. **318**(1–2): 167-175.

Joly, C., et al. (1999). "Residual solvent effect on the permeation properties of fluorinated polyimide films." Separation and Purification Technology **16**(1): 47-54.

Lam, C.-k., et al. (2005). "Effect of ultrasound sonication in nanoclay clusters of nanoclay/epoxy composites." Materials Letters **59**(11): 1369-1372.

Mohr, J. M., et al. (1991). "Gas transport properties of a series of poly(ether ketone) polymers." Polymer **32, 13**: 2387-2394

Recio, R., et al. (2007). "Gas separation of 6FDA–6FpDA membranes: Effect of the solvent on polymer surfaces and permselectivity." J. Membr. Sci. **293**(1–2): 22-28.

Ren, S., et al. (2005). "Surface modification of sulfonated poly(ether ether ketone) membranes using Nafion solution for direct methanol fuel cells." Journal of Membrane Science **247**(1–2): 59-63.

Rich, D. C., et al. (1997). "Effect of cure conditions on Probimide 32 polyamide-imide." Journal of Applied Polymer Science **63**(9): 1113-1126.

Santos, H. M., et al. (2009). The Power of Ultrasound. Ultrasound in Chemistry: Analytical Applications. J.-L. Capelo-Martinez. Weinheim, Wiley-VCH.

Shan, J., et al. (2006). "Sulfonated polyether ether ketone (PEEK-WC)/phosphotungstic acid composite: Preparation and characterization of the fuel cell membranes." Pure Appl. Chem **78**(9): 1781 - 1791.

Swier, S., et al. (2005). "Sulfonated poly(ether ketone ketone) ionomers as proton exchange membranes." Polymer Engineering & Science **45**(8): 1081-1091.

Wu, H.-L., et al. (2006). "Preparation and characterization of poly(ether sulfone)/sulfonated poly(ether ether ketone) blend membranes." European Polymer Journal **42**(7): 1688-1695.

Xing, D. M., et al. (2005). "Characterization of Sulfonated Poly(Ether Ether Ketone)/Polytetrafluoroethylene Composite Membranes for Fuel Cell Applications." Fuel Cells **5**(3): 406-411.

Xing, P., et al. (2004). "Synthesis and characterization of sulfonated poly(ether ether ketone) for proton exchange membranes." Journal of Membrane Science **229**(1–2): 95-106.

Ye, Y.-S., et al. (2012). "Water Soluble Polymers as Proton Exchange Membranes for Fuel Cells." Polymers **4**(2): 913-963.

Zhong, S., et al. (2008). "Modification of sulfonated poly(ether ether ketone) proton exchange membrane for reducing methanol crossover." Journal of Power Sources **180**(1): 23-28.

Fixed Plant Facility Siting – A Mega Project Lifecycle Framework

M.F. Haider
mech_haider@yahoo.com
mhaider@suncor.com

Prepared for Presentation at
American Institute of Chemical Engineers
2014 Spring Meeting
10th Global Congress on Process Safety
New Orleans, LA
March 30 – April 2, 2014

Fixed Plant Facility Siting – A Mega Project Lifecycle Framework

M.F. Haider
mhaider@suncor.com
mech_haider@yahoo.com

Keywords: Facility Siting, Project Management, Mega Project, Consequence Analysis (CA), Quantitative Risk Assessment (QRA), Adaptive, Change driven, Agile methodology

Abstract

Facility Siting based on tools such as Consequence Analysis (CA) and Quantitative Risk Assessment (QRA) is often times successfully applied for Brownfield operating unit projects, projects with relatively smaller footprints and for projects lasting relatively short time-periods. Challenge remains to apply and realize the full potential and value of Facility Siting for Mega-Projects. This is mostly due to project complexity and length – this paper will introduce a Management Framework to successfully apply CA and QRA as part of Facility Siting for Mega-Projects. The paper will also outline key strategies to ensure successful (i.e. value adding) implementation of facility siting, suggest a list of activities, associated sequence and pitfalls to avoid. Project Management team (e.g. Project Sponsor, Project Engineers, Managers) of both owner/operator and contractor/vendor may find this paper more useful as it provides a broad overview (i.e. when to initiate, how to plan, execute and close such assignment to maximize value) including real life examples of facility siting process over the project lifecycle.

1. Introduction

The driving force for process safety has been primarily based on catastrophic events [1]. With an increasing number of incidents, the process industry and governments started to take measures such as Facility Siting initiatives with hope to minimize loss of life and property. In the 1970s and 1980s, some of the world's most tragic industrial accidents took place. The Flixborough explosion in 1974 was reinforced by that of the Seveso tragedy in 1976 [2]. The Macondo blowout in the Gulf of Mexico on April 20, 2010, killed 11 employees and led to an uncontrolled oil spill lasting 87 days. This blowout was the most significant offshore incident in the US to-date. Finally, the Fukushima Daiichi nuclear plant incident in Japan in March 2011 drew the attention of the global process and power industries and encouraged them to incorporate natural disaster risks in a hazard analysis study [3]. As part of inherently safer design, many companies now-a-days apply CA and QRA as part of facility siting for their new projects. Due to the scarcity of information at the beginning, the studies in many cases start later in the project lifecycle for them to be effective and value adding. This paper will focus more on how to manage such initiatives for new Mega-Projects over the technical intricacies of the CA/QRA processes. The intention of the paper is to enhance the Body of Knowledge publicly available for management team in as much non technical languaga as possible to make their greenfiled Mega-Project initatives a success i.e. safer for people, environment and property.

So, what is different for industrial Mega-Projects?

- Industrial Megaprojects, as the name implies, are very large. To provide a simple and simply applied definition, we are defining a megaproject as any project with a total capital cost of more than $1 billion (U.S. dollars) as measured on January 1, 2003 [4].
- Most often engineering and construction activities run in parallel and not in sequence
- Multiple sub-projects under the 'Mega-Project umbrella' run in parallel
- Complex – design is almost 'never' complete in early stages – goes through many iterations
- Progressive elaboration is evident
- Scarcity of critical workforce with mega-project experience
- Scarcity of information – requires project team to make assumptions for CA/QRA
- Timeline: 2-5 years

2. Key Strategies

2.1 Facility Siting Charter

It is not only important or good to do but critical that the stakeholders need to identify, understand and agree on the purpose and value to conduct facility siting for their projects. The Author's personal experience over the years has been in order to maximize value of Facility Siting the project team has to start 'early' to develop project Charter and start with the end in mind. As can be seen from the following generalized Figure 1 that as the company starts to appraise opportunities, it is

Figure 1: Stages and Milestones to apply CA/QRA in a Mega-Project

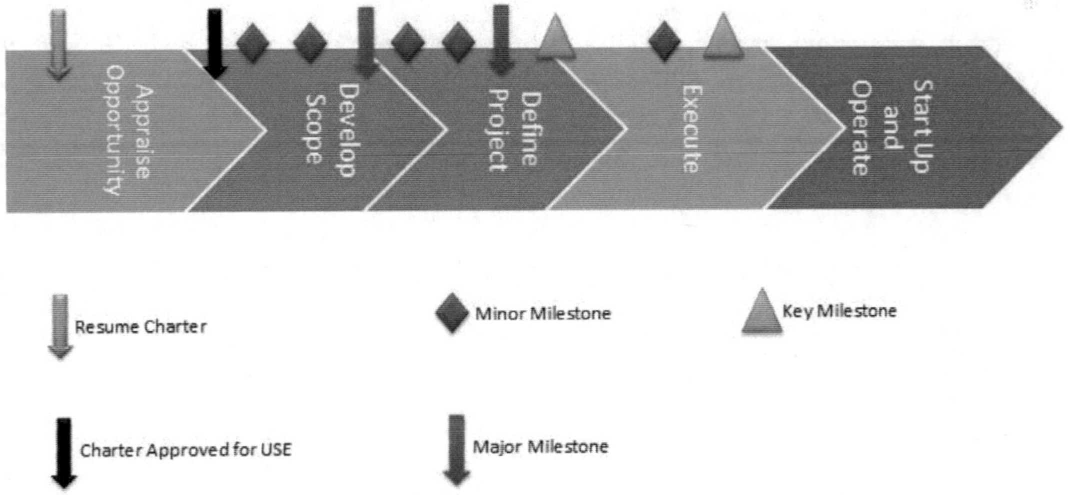

prudent to resume developing the Facility Siting Charter by bringing in the stakeholders together. Figure 1 shows a generalized project lifecycle for a typical Petro-Chemical Mega-Project. The

Figure also shows that the Key Deliverable is developed through a successive series of repeated cycles over the project life-cycle. The Milestones will be defined later in the paper. Part of starting with the end in mind is to clearly outlining ultimate goal/s, e.g. vision is to design a safer facility and utilize Mitigated QRA Report to formally document and communicate that facility is within the Risk Threshold or justification for accepting risk profile as is by Senior Management. The Vision in the Charter also provides opportunity to strike a balance between business goals and climbing the Risk Mountain – at each stage of increasing safety (decreasing risk). As part of striking a balance, the Project Engineer has to Identify options up-front if and/or when it becomes apparent that aggregate Residual Risk will not be within the Risk Threshold or acceptable so that team can put more effort in developing justification for either accepting or rejecting the risk profile by Management

Figure 2: Strike a balance between Business goal and Climbing the *expensive* Risk Mountain

2.2 Management Strategy

In order to successfully manage Facility Siting, CA and QRA for a Mega-Project, the stakeholders have to align on a management strategy. There are two primary management approaches for the purpose of managing such work:

1. Predictive or Plan driven
2. Adaptive or Change driven

The Author's past experience has been if the organization undertook very 'similar' projects in the near past (<5 years), critical experienced workforce is retained, Lessons Learned is readily available, and most importantly, the past project is still considered to be a success, it is advisable to use Plan driven approach. On the other hand, if the organization is introducing new technology, has limited past experience with similar project, involves relatively high number of stakeholders, capital cost > $1B, project life spans >2 years, it is recommended to utilize Change driven approach to maximize value. It is to be noted that it is not essential to meet all the suggested criteria mentioned above to pick one management strategy over the other.

2.2.1 Benefits of Adaptive or Change Driven Approach

In Adaptive/Change Driven Planning, project team intentionally repeats one or more project activities as the project team's understanding of the product increases ....Also known as ...agile methods .....intended to respond to high levels of change and ongoing stakeholder involvement [5]. The heart of Facility Siting involves making timely but informed decision. Some of the characteristics of this framework:

1. To add small incremental improvements that will deliver value to stakeholders
2. Milestones develop the product (e.g. Mitigated QRA Report) through a series of repeated cycles, while increments successively add to the functionality of the product
3. For each Milestone, a deliverable (read 'model') or set of deliverables will be completed. Future iterations may enhance those deliverables or create new ones. Each iteration incrementally builds the deliverables until the exit criteria for the phase are met, allowing the project team to incorporate feedback.

The biggest challenge with Mega-Project is scarcity of key information needed as inputs to develop scenarios and models. Some of the critical benefits, a project may gain by applying Change Driven approach:

1. The approach acts as Leading Indicator on Process Safety for Plant Layout - allows for multiple opportunities to make critical decisions on plant layout over the project life-cycle capitalizing on what is known at a given time. This lets the project team to use "what they know", to ensure "what they don't know" falls into the right place
2. Opportunity to make decisions over the project lifecycle not only works to save life but also helps with striking a balance between business needs and climbing the risk mountain
3. Aides design-development with progressive elaboration while producing final CA/QRA deliverable for the project

2.3 Execution Strategy

The main idea of Change Driven process starts with decomposing the known preliminary layout and process information into a set of requirements, work to be performed and their priority sequence, sometimes referred to as a *Product Backlog*.

It is advisable that the owner/operator forms a multi-use agreement (contract) with the chosen vendor. This way the plant areas under the Mega-Project umbrella can get their specific work done while the CA/QRA Project Team provides overall oversight and guidance.

At the beginning of an iteration, the team will work to determine how many of the highest priority items on the backlog list can be delivered within the next iteration. The time-length of each iteration should be less than 4 weeks. As shown in Figure 1, during "Develop Scope" and "Define Project", the length of each iteration could be even less than 2 weeks. This iterations (otherwise known as "Minor Milestone" as shown on Figure 1) will develop scenarios, assumptions and preliminary CA/QRA models. Assumptions are required due to lack of available quality information in the early stages of a project.

At the end of each iteration, the Minor Milestone/deliverable should be ready for review by the customers/stakeholders. This does not mean that the customer is required to accept delivery, just that the product should not include unfinished and/or incomplete features.

The sponsor and customer representatives should be continuously engaged with the other project stakeholders to provide feedback to Contractor on deliverables as they are created and to ensure that the *Product Backlog* reflects Sponsor/stakeholders' current needs. Below, you will find two examples using two different management strategies. For confidentiality purpose, the specific name/location is kept generic and few items were modified to suit.

2.3.1 Project X Facility Siting Using Change Driven Approach

A CA/QRA team was given the assignment to support design development during "Develop Scope" phase (Figure 1). The Sponsors already selected the site for the project. But the fixed plants were yet to be located on the site. The Vision for the initiative was for the QRA result to come under the organization Risk Threshold. Multiple EPC's were given to engineer different plant areas. The CA/QRA team selected a vendor to support them with defining the *Product Backlog* based on the preliminary Process Design Basis, intended product the facility is expected to sell, the feedstock necessary etc. The CA/QRA team had prioritized the Product Backlog based on hazard level of feedstock, preliminary PHA schedule and project schedule. This also involved obtaining feedback from different Process Engineering and Operations teams to collect necessary inputs to put together the assumptions and scenarios to be modeled within the 1st iteration.

Out of 25+ scenarios, only 7 perceived to be the most hazardous were modeled for CA that gave ample input (e.g. explosion, toxic and fire contours) for the CA/QRA team to work with the stakeholders to recommend location for specific process units on the given site. Note that, CA/QRA team integrated the potential consequence to understand local incident and global consequence events. At this time QRA was not performed. This was used as input for minor modification of preliminary plant layout. It is to be noted that the site plan was still preliminary after the first iteration but assisted different EPC accountable for their specific area to recognize where to locate the specific plants i.e. finalized key plans for the overall site plan.

With further design-development, more refined information became available from each EPC. This new information led to 130+ scenarios and was used for further CA/QRA iterations that created either new model and/or embellished the past ones. Each EPC was accountable for facility siting within their boundary while the CA/QRA team provided technical oversight and integration support for the overall project. The Sponsor/stakeholders were engaged on a regular basis to ensure that the Product Backlog reflects future modeling requirements to properly site facilities e.g. Fire Water Pump house, Tank Farms, site Medical Centre etc. Early engagement of CA/QRA team and higher management willingness to incorporate their feedback with the progressive elaboration of the project led to QRA result meeting the organization risk threshold and ultimately safer facility. Changes suggested based on CA/QRA work in earlier stage (i.e. Develop Scope, Define Project) were easier to incorporate into the design.

2.3.2 Project Y Facility Siting Using Plan Driven Approach

The sponsors for project Y selected the site for the project and decided to hold on to not perform CA/QRA work with the progressive elaboration of the project activities. This was mostly decided based on the fact that not all information was available to properly conduct the 'studies'. Fundamentally, the sponsors looked CA/QRA as deliverables requiring a more 'complete' plan. Once the Develop Scope (as referred in Figure 1) is almost complete, the sponsors decided to engage CA/QRA team and a vendor. By this time, the project had significantly high number of stakeholders. Many of the stakeholders were not very familiar with CA/QRA and the value it could bring to them. There was not much time to schedule and train the stakeholders as well to bring them up to speed. As a result, the stakeholders were unable to align on a Vision. The Vendor was asked to provide explosion, toxic and fire contour maps for specific scenarios and locations that were almost 'fixed' on the site plan at that point in time. The resultant contour maps showed significant consequence on nearby facility with over 700+ people. It was also noted that the plant area consuming the most hazardous substance would impact the nearby facilities significantly from consequence perspective. It should be noted that the result was only based on a selected few scenarios. By the time the project reached the "Execution" phase, more scenarios were added as part of the scope. But it was already too late to incorporate most of the recommended changes resulting from CA/QRA work. The project went through several area by area risk reduction workshops only to accept (by the company executives per organization risk matrix) the aggregate residual risk level which was also found to be not within the organization risk threshold.

2.3.3 Importance of Starting Early

Figure 3: Impact of variable based on project time [5]

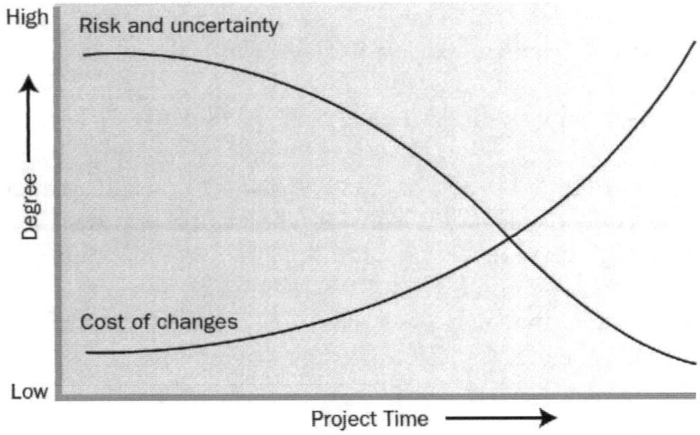

The above two examples highlight the importance of starting CA/QRA work early enough to realize and maximize the value CA/QRA brings to a Mega-Project. The high-level view provides common frame of reference for comparing projects for CA/QRA initiative, even if the projects were dissimilar. Change driven methodology allows for applying the known information at early stage in a way that supports laying out plant facilities to reduce risk and thus maximize potential value. While these characteristics remain present to some extent in almost all project life cycles, they are not always present to the same degree [5]. Change Driven framework, in particular, keeps stakeholder influences higher and the costs of changes lower especially in the early stages of the life cycle than in predictive/plan life cycles. The Figure 3 highlights a well-known fact of project management that even though risk and uncertainty remains high in the early stages of a project, cost of incorporating changes remains much lower. This translates into a key opportunity for project stakeholders to align on laying out the key areas on a site plan intelligently to reduce risk and maximize value by applying tools such as CA and QRA.

2.4 Key Decision Points

Facility Siting, from a management point of view, is all about bringing the stakeholders closer and find the 'right' spot for the facilities. Finding the 'right' spots requires making timely decisions. The challenge has always been to make timely decisions as the stakeholders more often find themselves forced to make those decisions based on information that are frequently of poor quality and at times uncertain, especially in the early stages of a project. At times, due to the scarcity of information, stakeholders rely on their past experience to make those decisions on plant layout. As a result, depending on the individual experience, two different teams for a project area may come up with two very different plant layout. CA/QRA work, if started early enough, will also depend on input that most often requires embellishment through assumptions to create needed scenarios and models. Even with this limitation, CA/QRA provides a key 'input' in allocating real estate for key facilities on a plant site at early stage of a project lifecycle. With each iteration and deliverable 'Change driven' methodology allows for multiple opportunities to make critical decisions on plant layout over the project life-cycle. It allows the stakeholders to decide when to 'sharpen the pencil' (e.g. PHAST to Computational Fluid Dynamics model), relocate facility or accept the consequence and/or risk, 'pull the plug' on further modeling/refinement and start developing justification, if required, to recommend management to accept/reject the aggregate risk profile.

2.5 Alignment with Process Hazard Analysis (PHA) Planning

CA/QRA results are a useful input to PHA (e.g. HAZOP, LOPA) activities. This is particularly true as projects move forward after establishing the plant layout on a high level and starts to conduct "Define Project" (Figure 1). PHA activities may also result in recommendations (i.e. new scenarios) that goes into the *Product Backlog* and requires project team to develop new models or refine old ones. Mega-projects have multiple sub-projects running in parallel most of the time, having an Integrated Schedule for all the reviews is a good strategy so that the right stakeholders are booked for the sessions. The integrated PHA schedule also supports prioritization of *Product Backlog* for CA/QRA activities over the lifecycle of the project.

2.6 Integration Strategy

It is critical for a Mega-Project involving multiple plant areas/silos to understand the BIG picture so that site risk can be dealt with on an aggregate basis. There is strong possibility of one area to impact other plants in the vicinity. It is recommended to review integration aspect and aggregated results on a regular basis. The following Figure 4 shows that the origin of an incident may reside in one particular plant area, while it may impact other areas in the vicinity.

Figure 4: Consequence of an incident in one area impacting multiple areas

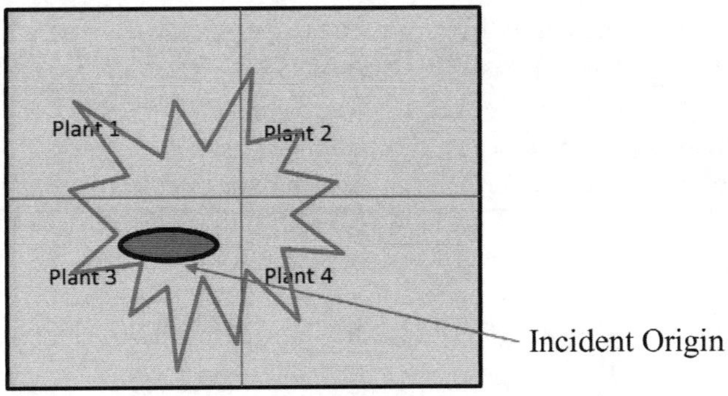

Recommended Strategy:

- Setup Strong Center - Task Force Single Point of Contact (SPOC)
- Collaborate – SPOC will promote Collaboration among Plants/Areas

3. List of Activities (suggested)

3.1 Overall Sequence

The CA/QRA work typically should start with the approved Charter as shown on Figure 1. A project Task Force i.e. SPOC is formed comprising of Project Engineer, Operations rep, Process rep and CA/QRA Technical Expert. The SPOC is ultimately accountable to produce the CA/QRA deliverable studies but also acts to support the design development and site plan finalization efforts in the early stages of the project.

The team collects 'input' such as Preliminary site Layout and design information, works with area process and operations team to identify scenarios, develop assumptions for known-unknowns and unknown-unknowns, develop modeling 'Product backlog', prioritize backlog based on PHA schedule and produce modeling output (i.e. deliverable) for stakeholder use. The stakeholder usually has the option of accepting or rejecting consequence. If the result of an iteration is acceptable within area causing the incident – communicate to outside stakeholders to ensure there is no cross boundary impact. If the result of the iteration is NOT acceptable to outside of the area – use the result as input to PHA for the areas to mitigate. Collaboration recommended between origin area and impacted area/s to identify appropriate mitigations. There is also a possibility that the consequence resulting from the iteration is not acceptable to the area causing the incident, in that case the area will initiate a PHA to mitigate the consequence. As part of developing mitigations, consider the following:

- Develop QRA if appropriate
- Perform Building Assessments for Critical Buildings
- Apply Cost-Benefit analysis to optimize risks/opportunities

- Determine Safe Haven locations and their characteristics

If Mitigation is not possible from Business perspective – perform As Low As Reasonably Practicable (ALARP) justification for Senior Management to formally accept the posed risk

- To apply mitigations on a site wide basis, review Milestone deliverables by taking on a BIG picture perspective
- Future iterations may enhance those deliverables or create new ones as new information is available. Each iteration incrementally builds the deliverables to produce the Mitigated CA/QRA report to support Facility Siting for the stage

The overall effort is expected to produce a maximum of two single reports for the overall site:

1. Mitigated version of QRA Report
2. and Consolidated CA Report

It is recommended to develop as-built, pre-start up CA/QRA reports prior to start-up. Table 1 and 2 outlines typical activities performed by SPOC and plant areas respectively once key process areas are established/finalized on Site Plan. Please note that the activities are project dependent and may warrant modifications to suit specific situation.

Table 1. SPOC Sequence of Activities to conduct Facility Siting

Initiative	Key Steps
Setup Facility Siting Team	Sponsor to support to setup a team (i.e. Project Engineer, Technical Expert, Operation, Process representative) to act as SPOC
Integration	Provide functional (technical and management) guidance to Areas
	Assess cross boundary impact/opportunity
	Communicate incremental outputs after each Area PHA/Risk Reduction
IFA versions (prior to "Define Project" stage)	Consolidated CA Report
	Mitigated and Integrated QRA Report
IFD CA/QRA (prior to "Execution" stage)	Re-validate IFA CA/QRA post PHA and vendor PHA
As-built versions	Post-oil version

3.2 Typical Area Activities

Table 2. Area (Typical) Sequence of Activities to conduct Facility Siting

Initiative	Key Steps
Plant 1 (Typical)	Perform Modeling and Calculations
	Conduct Risk Reduction Workshop (if appropriate) – Apply mitigations
	Incremental output for use based on reduced risk level
	Assess Buildings/facilities
	Identify Safe Haven/shelter Locations (if required)
	Estimate Area Project Scope impact
	Perform area scope updates/changes e.g. passive fire protection requirement
	Carry out QRA (if appropriate)
	Update Product Backlog
	Develop/update contract to support area specific activities

4. Pitfalls to Avoid

The following observations were noted by the Author during his time to manage Facility Siting scope:

1. Facility Siting does not replace good and common sense engineering practice e.g. locating a plant that uses high volume of hazardous material will significantly increase risks to other plants/areas (e.g. camp) within the vicinity
2. Evaluate effectiveness of 'Cost Reduction' philosophies as opposed to added risk e.g. different modularization techniques may decrease the plant footprint by reducing pipe lengths but makes the plant more compact and likely to increase severity of explosion consequence
3. Avoid developing models with hope to realize risk reduction
4. Sharpening the pencil for models most likely may not reduce risk
5. Avoid jumping into CFD or other similar methods without evaluating cost-benefit for undertaking such effort
6. Utilization of QRA results is much more controversial than the methodology. Watson [6] has suggested that QRA should be considered as an argument, rather than a declaration of truth. In this view – focus on accuracy over precision
7. CA/QRA Task Force should be part of Management of Change (MOC) process
8. Evaluate carefully before locating facilities e.g. tank farms close to main entrance/egress as part of Facility Siting

9. Keep the focus on VISION for the overall effort to avoid scope creep
10. Note the difference between Facility Siting as a Deliverable and tool for design-development activities

5. Conclusion

This paper presented Adaptive/Change driven framework to manage initiatives like facility siting/CA/QRA as part of Mega-Projects. Change driven framework does not mean "lack of planning or no planning". As demonstrated in the paper, this methodology starts with a vision and then develops the ultimate product both iteratively and incrementally when it is possible to define small incremental improvements that will deliver value to stakeholders. This allows making key decisions on laying out the facilities in early stages of a project rather than waiting for design to develop to a significant extent. Often times, project team wait too long for design to mature enough before applying CA/QRA to validate consequence impact and/or risk level. Most often, especially for Mega Projects, this leads to delays in executing the CA/QRA work properly at the right time to influence site layout. This delay in execution leads to scenarios when scope changes cannot be incorporated anymore. This translates into value erosion as project was unable to reap the full benefit of Facility Siting. The management framework recommended in this paper i.e. Change driven/adaptive framework allows for the stakeholders to start early in the project lifecycle to apply CA/QRA in laying out the plant facilities. The key is to start EARLY but START with the END in mind. Alignment among stakeholder on a VISION up-front is also important as this is expected to save life while striking a balance between business needs and climbing the Risk Mountain. Vision also allows avoiding scope creep while applying Adaptive/Change driven approach. It is highly recommended that stakeholders focus more on accuracy over precision. In the early stages of project lifecycle, quality information is scarce. Change driven/Adaptive framework gives you as the project stakeholder the opportunity to use "what you know" intelligently to ensure "what you don't know" falls into the right place.

6. References

[1] M. Mannan, A. Chowdhury and O. Reyes-Valdes, "A portrait of process safety: From its start to present day," College Station, 2012.

[2] M. Mannan, Lees' Loss Prevention in the Process Industries, Volumes 1–3 (3rd Edition), , 2005, Elsevier, 2005.

[3] K. McAndrews, "Consequences of Macondo: A Summary of Recently Proposed and Enacted Changes to US Offshore Drilling Safety and Environmental Regulation," in *Society of Petroleum Engineers, Americas E&P Health, Safety, Security and Environmental Conference*, 2012.

[4] E. W. Merrow, Industrial Megaprojects Concepts, Strategies and Practices for Success, New Jersey.: John Wiley & Sons, Inc., 2011.

[5] A Guide to the Project Management Body of Knowledge (PMBOK® Guide) – Fifth Edition, Newtown Square: Project Management Institute, Inc., 2013.

[6] S. Watson, "The Meaning of Probability in Probabilistic Risk Assessment," *Reliability Engineering and System Safety,* vol. 45, pp. 261-269, 1994.

348480 Global Energy and Transportation and Mobil Oil

Abstract:

Biofuels, Computational Methods, Engineering Economics, Polymers, Thermodynamics. Polymers-plastic, Thermodynamics- oil, Engineering Economics – the money it takes to finish the method of Global Engineering Research, Computational Methods – numerical input of analytical data, Biofuels – the transport fluids that utilized in more than one method from industrial usage to everyday living which enters the atmosphere. Exxon has completed a $600 million dollar deal for production of algae biofuel, amongst Mobil Oil has performed major research Global Energy and the effects of oil. Reading a most recent article on Mobil website.

348748 Characterization of Iron Phthalocyanine As the Active Material for Lithium Batteries

Abstract:

Results on the determination of the lithium-ion diffusion coefficient in the iron phthalocyanine (FePc) active material as a function of intercalation level, using a modified electrochemical impedance spectroscopy technique, will be presented. Also, developed thermodynamic functions such as the Gibbs free energy, entropy, and enthalpy changes for the lithium insertion reaction with FePc, derived from cell open-circuit voltage measurements as a function of intercalation level, are presented. Also presented are estimates of the maximum amount of lithium that can be intercalated into the FePc based on constant volume and temperature conditions.

349086 Design of a Free-Fall Reactor for Fast Pyrolysis of Waste Plastics

Abstract:

Thermal pyrolysis has earned significant reputation in the recovery of valuable chemicals, especially petrochemical based fuels from various materials, including biomass and plastics. Of many available technologies, *fast* pyrolysis was preferred in his study owing to its minimal heat and mass transfer limitations and better control on process conditions. The focus of this study is towards efficient conversion of waste plastics into gaseous products consisting primarily of alkanes, alkenes and aromatics. Lab scale experiments were performed using a micro reactor to determine favorable conditions, including reaction temperature and particle size that would result in maximum yield of ethylene. Using these results, a simple industrial size radiatively heated free-fall reactor was designed for the purpose of pyrolyzing waste plastics. The goal is to design a *fast* pyrolysis reactor that improves upon inefficiencies in traditional reactors while resulting in high yield of ethylene.

349732 Biosorptive Dehydration of Ethanol/Water Azeotropes Using Compound Starch-Based Adsorbent

Abstract:

Abstract: Recently, the interest in the production of bioethanol has greatly increased as a result of the increasing fossil fuel prices and associated environmental issues. Bioethanol is made from either grains or biomass and the broth contains 6 to 12 percent ethanol. In addition, water removal represents a serious problem as ethanol form azeotrope with water at atmospheric pressure. Distillation to water-free ethanol consumes 50 to 80 percent of energy used in the fermentation ethanol manufacturing process. Various techniques have been developed to break the azeotropic point, such as salting-out method, azeotropic distillation, extractive distillation, reactive distillation, adsorptive distillation and pervaporation. Among the methods, the adsorptive option is the most attractive, especially the biosorption, which has distinct advantages over conventional methods as it is environmentally friendly, highly selective, easily available, easy to operate, cost effective and reusable in repeated cycles in the treatment of aqueous organics.

In this work, a specially formulated compound starch-based adsorbent (CSA), which consists of corn, sweet potato and foaming agent, was developed. As an exploratory study, this research was focused on the investigation of CSA for biosorptive dehydration of aqueous ethanol and optimization of conditions for its maximum adsorption under various conditions of vapor feed flow rate, bed temperature and bed height through an orthogonal design method. Furthermore, to investigate the adsorption mechanism, dynamic saturated adsorbance of CSA was studied and characterization of the adsorbent before and after adsorption, as well as after regeneration, was performed.

In the first step, the net retention time and separation factor of ethanol and water were obtained using inverse gas chromatography (IGC), the results indicated that the net retention time of water on CSA is significantly longer than that of ethanol which attested the feasibility of this adsorptive separation purpose and low temperature could be propitious to the process which was consistent with the traditional theory of adsorption.

Then, through a batch adsorption experiment, the optimum adsorption condition was determined under vapor feed flow rate, bed temperature and bed height by orthogonal design method, which indicated that the optimization condition for ethanol was at vapor feed flow rate of 1.1g/min, bed temperature of 83 and bed height of 25cm. The packing density was 445kg/m^3. The residence time is determined by vapor feed flow rate and bed height. The residence time is more influenced than bed temperature. The longer the residence time, the more the output is. But it cannot be ignored that the adsorption selectivity may decreases. Low bed temperature is benefited to the adsorption process which agrees with the result of IGC. Five cycles experiments of adsorption-regeneration were carried out at the optimum adsorption and regeneration condition, the results showed that there is practically no change in the regeneration efficiency of CSA after the cyclic process adsorption–regeneration. It is due to the properly chosen regeneration conditions assuring complete desorption of the water adsorbed without changes in the CSA structure. This unchanged adsorption capacity after a number of cycles demonstrates reusability of the adsorbent.

To study the adsorption mechanism, experiment was conducted to study the dynamic saturated adsorbance of CSA. The water adsorption capacity of CSA has been found to be similar to some molecular sieves, which is about 0.15g/g. Besides, field emission scanning electron microscopy (FESEM) and mercury porosimetry (MP) were applied to investigate the change of surface morphology and microstructure of CSA before and after the adsorption, as well as after regeneration, as shown in Fig.1 and Fig.2. The FESEM images showed that compared with the particles before the adsorption, they became plump and the void spaces became smaller after adsorption, which indicated that the CSA was saturated with water. However, the CSA particles after regeneration became wizened like the nature particles, which demonstrated that almost all the water had been desorbed from CSA. The pore size distribution (PSD) indicated that compared with the PSD curve before adsorption, the peak at 6000nm disappeared and the peak at 17000nm became remarkably smaller after adsorption, which manifested that CSA is mainly composed of macropores which play a major role in the adsorption. However, the peaks at 6000 and 17000nm reappeared after regeneration, and the curve looked exactly the same with that before adsorption, which further demonstrated that CSA after regeneration can be totally regenerated and recovered to the original structure.

These results indicated that the starch-based adsorbent can be considered as an effective and low-cost bio-sorbent to dehydrate alcohol.

349781 Purification of 2-Amino-1-Phenylethanol Enantiomers By a New Technique Combining Distillation and Crystallization

Abstract:

Stripping crystallization (SC) is a new separation technique which combines distillation and crystallization. Basically, SC is operated at a triple-point condition, in which the liquid is simultaneously vaporized and crystallized due to the three-phase equilibrium. In essence, the process is continued until the liquid phase is completely eliminated and only the pure solid crystals remain in the feed. Thus, crystal washing is not required since only pure solid crystals remain in the feed and no impurities are adhered on the crystal surfaces at the end of SC. In the previous research, SC has been successfully applied to separate p-xylene from the mixed xylenes by Shiau et al. (2005, 2008).

Phenylethanolamine, also known as 2-amino-1-phenylethanol (2-APE), is an important basic skeleton in many adrenergic drugs and is often used as ingredients of cold medicines and anorectics. As the biological activity and pharmacokinetics of the enantiomers of a chiral drug can be different, the objective of this research is to apply SC to purify R-2-APE from the chiral 2-APE mixture. The experiments show that no crystallization occurs during SC without seeding due to subcooling of the liquid enantiomers. On the other hand, seeding can eliminate subcooling during SC, leading to the R-2-APE crystals gradually formed preferentially on the seeds. Thus, seeding with magnetic stirring can be successfully employed to purify R-2-APE from the 2-APE enantiomers.

1. Shiau, L. D., Wen, C. C. and Lin, B. S. Separation and purification of p-xylene from the mixture of m-xylene and p-xylene by distillative freezing, *Ind. Eng. Chem. Res*, **44(7)**, 2258-2265, 2005.

2. Shiau, L. D., Wen, C. C. and Lin, B. S. Separation of p-xylene from the multicomponent xylene system by stripping

350841 Fabrication of Low Cost Insulating Material from Kaolin Clay for Construction Purposes

Abstract:

A new low cost, thermal insulator, construction mineral polymeric material was synthesized from local Saudi Arabia resources using Inorganic Mineral Polymerization technique (MiP). In this technique, the mineral polymer is produced from the following components: clay mineral rich in koalinite, sand, sodium hydroxide solution, and water. Inorganic polymerization experiments are conducted by mixing known amounts of clay minerals and sand in a variable speed stainless-steel mixer for some period of times until a homogenous solid mixture is obtained. Sodium hydroxide solution is added to the mixture and the polymerization reaction takes place at different mixing speeds and times. The produced mixture is divided into a number of samples of known weights. Each sample is moulded in a stainless-steel cylinder at a medium pressure. Then the moulded samples are cured in an oven for 24 hours. The curing temperature was 80°C. The mineral polymerization conditions such as the ratio of clay to sand filler, particle size of the clay, particle size of the sand, concentration and amount of sodium hydroxide, speed and time of mixing were optimized to obtain stable and hard mineral polymer. Compressive strength for different specimens of fabricated mineral polymers was tested under dry and wet conditions. Values of compressive strength with values ranging from 11.5 MPa to 27.9 MPa were obtained.

351141 Thermodynamic Study of Binary PAH (Anthracene + Phenanthrene) Solid Mixtures

Abstract:

Polycyclic aromatic hydrocarbons (PAHs) are common components of many materials, such as petroleum and various types of tars. They are generally present in mixtures, occurring both naturally and as byproducts of fuel processing operations. It is important to understand the thermodynamic properties of such mixtures in order to better understand and predict their behavior (i.e., fate and transport) in the environment and in industrial operations. The Knudsen effusion technique was combined with gas chromatography and differential scanning calorimetry to measure the vapor pressure, composition, and phase behavior of binary anthracene and phenanthrene mixtures at near ambient temperatures. Mixtures of anthracene and phenanthrene exhibited non-ideal mixture behavior: In mixtures with less than 80% phenanthrene, this more volatile component preferentially sublimed and the vapor pressures decreased as the mixtures became enriched in anthracene. The vapor pressure data do not follow Raoult's law (that is, ideal mixture) predictions. Mixtures with high phenanthrene content (> 80%) showed azeotrope like behavior, in that anthracene and phenanthrene sublimed at nearly constant vapor pressure and molecular ratio. These results confirm that the behavior of few-component PAH mixtures does not follow ideal mixture models. Hence, as tars and petroleum (and potentially other organic compound-containing mixtures) weather to few component mixtures over time, complicated, non-ideal phase behavior might be observed.

351160 How Confident Are You That Your Major Accident Risks Are Under Control?

Abstract:

Major accidents in the process industries including those at Texas City , Buncefield UK and in the Gulf of Mexico have highlighted the critical role played by senior managers and executives. Effective leadership is essential if companies are to develop a positive safety culture that remains constantly vigilant towards the risk of a major process safety incident. In the UK the HSE has made leadership a key priority for the high hazard industries and there is a cross-industry initiative to develop and deliver training focussed towards board members responsible for process safety performance.

This paper explores the leadership failings that contributed to recent major accidents, and the cost to both the company reputation and the individuals involved. Key principles are discussed that can help improve leadership in the major hazard industries. These include; attitudes and decisions of senior manager affecting the investment strategy and the safety culture of the organisation, reinforcing the importance of safety by personal example, thorough understanding of major accident hazards and key risk control systems, investigating process safety incident and near misses to find the underlying causes, developing world class safety management systems, and identifying weaknesses in these systems using targeted performance indicators.

The author bases this paper on the development of a training standard for 'Process Safety Leadership' involving a cross-industry 'expert panel'. Training of senior executives in compliance with this standard has been carried out from late 2011 and feedback from these courses and the key requirements for effective leadership will be discussed.

351236 Sil Determination: Shortcomings with the Use of LOPA

Abstract:

Proposed Topic Area: Managing Risks

For a number of years, industry has been trying to apply the principles outlined in IEC 61508. In the process sector, this has resulted in a focus on the requirements of IEC 61511 (2003) and ANSI/ISA 84.00.01 (2004). These are risk based standards. They involve the identification of potential hazardous events, the causes that could lead to each event and the necessary risk reduction for each event. Where the risk reduction involves an instrumented function (e.g. a trip or an interlock), the required performance of the instrumented function can described in terms of the average probability of failure on demand. In the standards, the performance for such instrumented functions is assigned to one of four Safety Integrity Levels. The assessment of what performance level is needed in respect to a specific hazardous event is referred to as SIL Determination. In the field of SIL Determination, IEC 61511 provides examples of a number of different approaches – these are shown in IEC 61511 Part 3. These include Event Trees, Risk Graphs, Safety Matrices, Fault Tree Analysis and Layer of Protection Analysis (LOPA).

In the first few years, following the publication of IEC 61508 Edition 1 in 2000, industry generally embraced the use of Risk Graphs as the method of choice. However, over the subsequent few years, many companies using this method found that it was (a) giving higher safety integrity level targets than really needed and (b) was less flexible than other methods and consequently more difficult to use in many real situations. This has resulted in a general move to other methods and in particular to the use of Layer of Protection Analysis. Indeed, LOPA may now be the predominant "Method of Choice" for SIL Determination. As a method, LOPA gives the impression of simplicity and appears quite straightforward. It almost invites anyone with a minimum of understanding of the method to be a practitioner.

However, there are a number of shortcomings in the use of the LOPA method that have become evident through its use across the globe.

This paper will review the application of LOPA and provide insights into what may be seen as the proper use of the method and its core principles. It will also highlight various shortcomings in use that have been found during expert reviews carried out of a number of existing studies. The paper will be of benefit both to those who are new to the LOPA method, but also of value to those who have been using the method for some time.

Shale Oil and Gas – Wellhead to Fuel

Dr. Lawrence Kremer - Baker Hughes
AIChE - Hilton New Orleans Riverside
April 1, 2014

14AIChE Spring Meeting, New Orleans, LA

BAKER HUGHES

Outline Unconventional Shale Oil

- Properties of shale oils
- Problems
  - Transportation
  - Refining

14AIChE Spring Meeting, New Orleans, LA

BAKER HUGHES

General Characteristics of Shale Oils

- API gravity typically varies 30-55+
- Very paraffinic
- Low sulfur
- Can have hydrogen sulfide and mercaptans
- Low TAN
- Very low asphaltenes
- Light crude with very little residual material
- Many samples show great quality variation

Factor	Bakken	WCS
API	39.7	21.6
S%	0.25	3.44
Light ends	2.6%	
LSR	3.5%	3%
Light Nap	12.4	3.0
Heavy Nap	9.7	5.0
Jet	10.5	10.0
LFO	27.6	12.0
Gasoil	27.2	37.0
VacTar	6.7	29.8

14 AIChE Spring Meeting, New Orleans, LA

BAKER HUGHES

Shale Oil Property Comparison

Parameter	Eagle Ford	Bakken
API	52	40.8
TAN (g KOH/g)	< 0.05	0.09
Sulfur (wt %)	<0.2	0.304
Asphaltene (wt %)	0.1	0.41
Resin (wt %)	1.6	4.95
Filterable Solids (PTB)	225	76

14 AIChE Spring Meeting, New Orleans, LA

BAKER HUGHES

Utica Shale Crude - Three Phase

14 AIChE Spring Meeting, New Orleans, LA

BAKER HUGHES

Eagle Ford Crude Samples

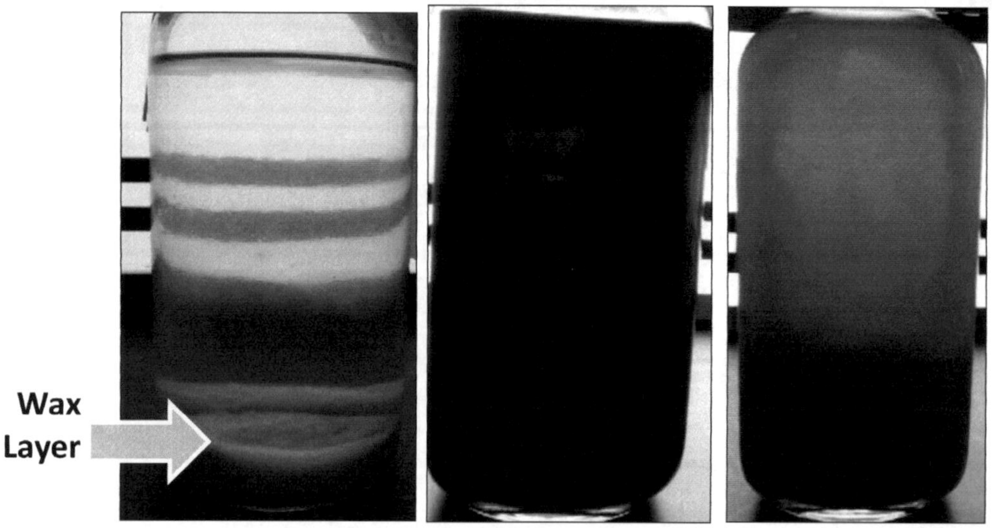

14 AIChE Spring Meeting, New Orleans, LA

BAKER HUGHES

Analytical Breakdown

Value	Yellow	Red	Black
API	55.0	44.6	52.3
TAN	<0.05	0.07	<0.05
Sulfur	<0.20%	<0.2%	<0.2%
Na	1.0 ppm	1.6 ppm	1.6 ppm
K	0.3 ppm	0.4 ppm	0.5 ppm
Mg	3.4 ppm	2.9 ppm	3.0 ppm
Ca	2.6 ppm	2.8 ppm	3.8 ppm
Asphaltene	0%	0%	0.1%
Resin	0.5%	3.2%	1.6%
Filterable solids	176 PTB	295 PTB	225 PTB
P	5.3 ppm	4.6 ppm	5.0 ppm

14AIChE Spring Meeting, New Orleans, LA

BAKER HUGHES

Micrograph of Sludge

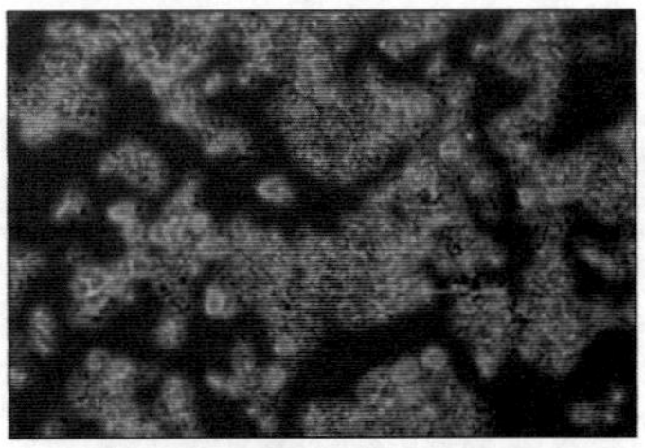

100 X

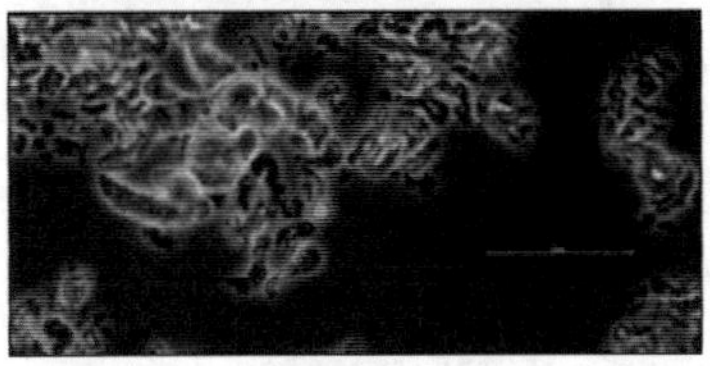

400 X

14AIChE Spring Meeting, New Orleans, LA

BAKER HUGHES

Paraffin Analysis of Sample

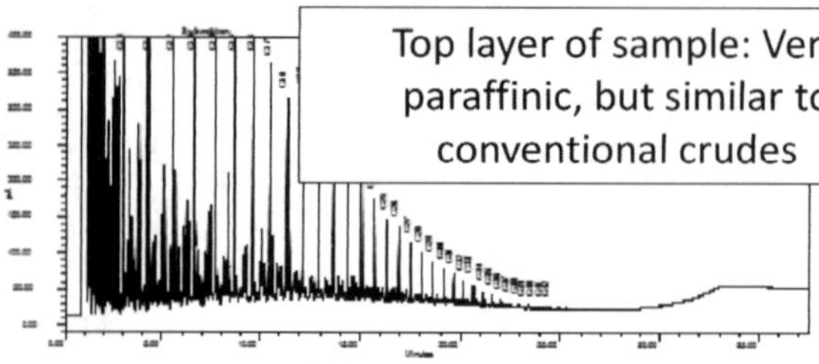

Top layer of sample: Very paraffinic, but similar to conventional crudes

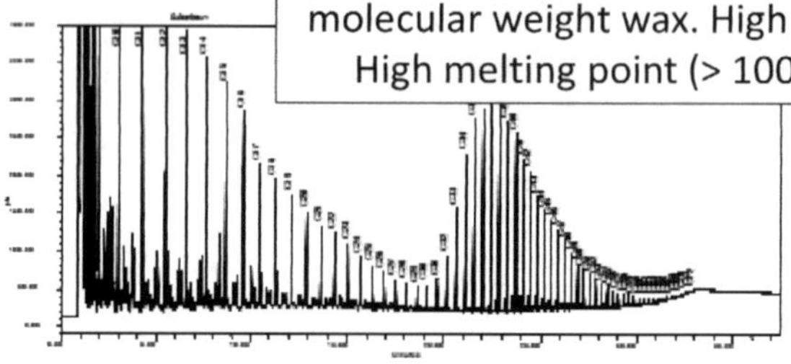

Bottom layer: Considerable high molecular weight wax. High MW = High melting point (> 100 °C)

14 AIChE
Spring Meeting, New Orleans, LA

BAKER HUGHES

Transport of Shale Production

- Infrastructure does not exist to meet needs
  - Most shale oil moved by rail or trucks
  - Limited pipelines expanding to meet transportation needs
- Unit trains to East Coast, West Coast, and Gulf Coast
- Problems:
  - Tankers and barges with heavy wax deposits ROB
  - Tight emulsions that look like oil or white condensate but high BS&W
  - Cars not uniformly treated for H_2S
  - Corrosion in rail cars

14 AIChE
Spring Meeting, New Orleans, LA

BAKER HUGHES

Viscosity of ROB Before Treatment

ROB from untreated compartment

14 AIChE
Spring Meeting, New Orleans, LA

BAKER HUGHES

Eagle Ford Shale Oil Wax Dispersion Testing

Several products were effective in dispersing the paraffin wax in an Eagle Ford shale oil sample at ambient temperature for over a month.

14 AIChE
Spring Meeting, New Orleans, LA

BAKER HUGHES

Product 'A' Effect on Wax Deposition (Utica)

Blank 200 ppm 400 ppm

14 AIChE
Spring Meeting, New Orleans, LA

BAKER HUGHES

Refinery Issues

- Crude Blending
- Fouling
- Corrosion
- Product quality

14 AIChE
Spring Meeting, New Orleans, LA

BAKER HUGHES

Crude Unit Fouling

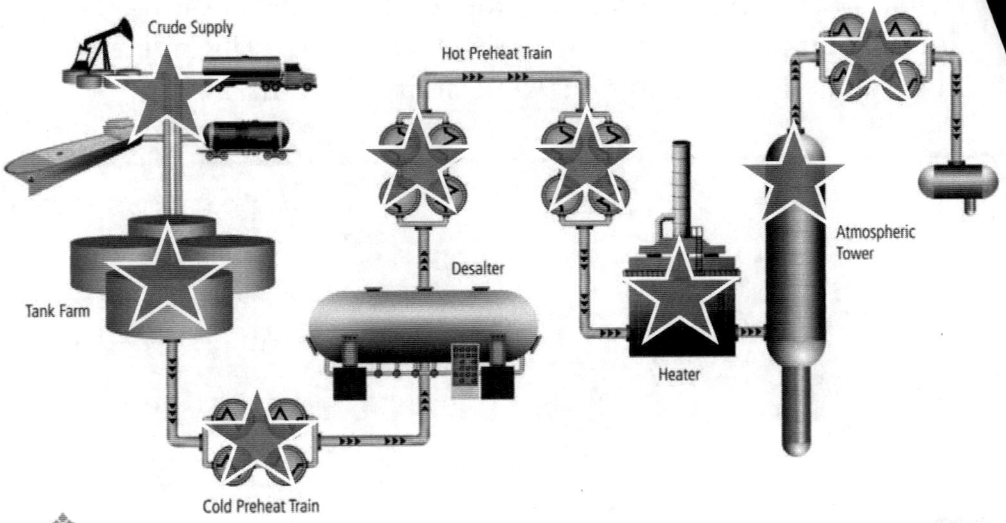

★ = Susceptible to Fouling

14 AIChE Spring Meeting, New Orleans, LA

BAKER HUGHES

Asphaltene Precipitation

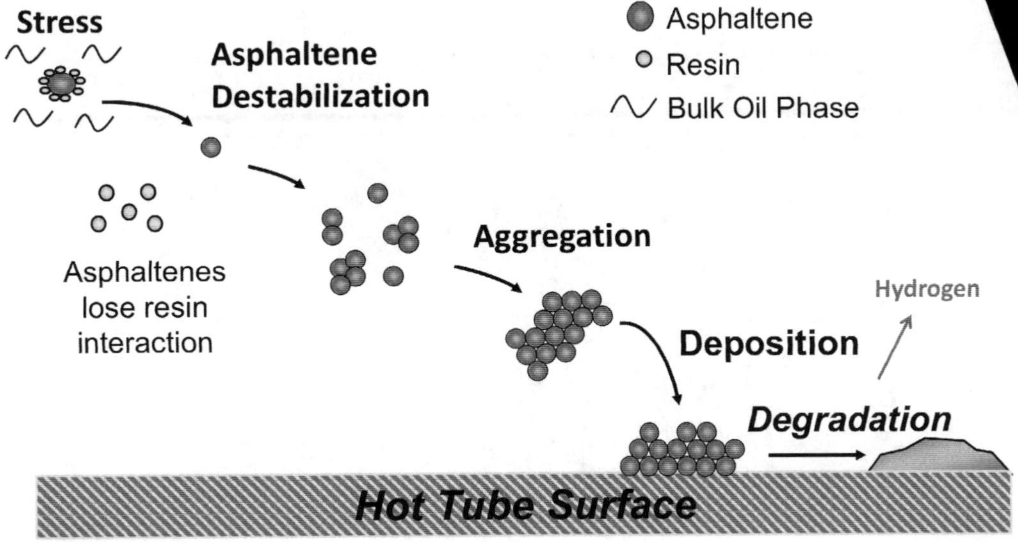

14 AIChE Spring Meeting, New Orleans, LA

BAKER HUGHES

Asphaltene Stability Index Test

Field ASIT services™ technology measures the onset of the flocculation of the asphaltenes with high accuracy by inducing the asphaltene precipitation via titration with a paraffinic solvent.

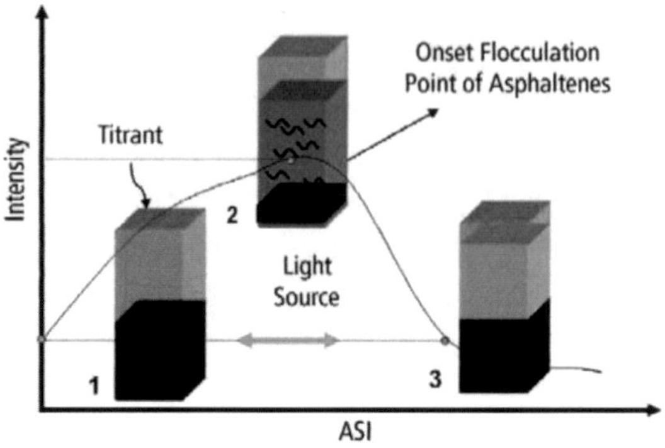

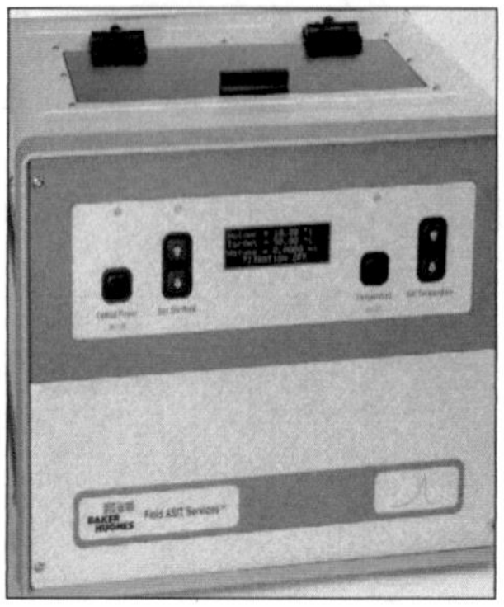

Field ASIT services is a trademark of Baker Hughes Incorporated

14 AIChE Spring Meeting, New Orleans, LA

BAKER HUGHES

ASI Shows Eagle Ford Can Destabilize Crude

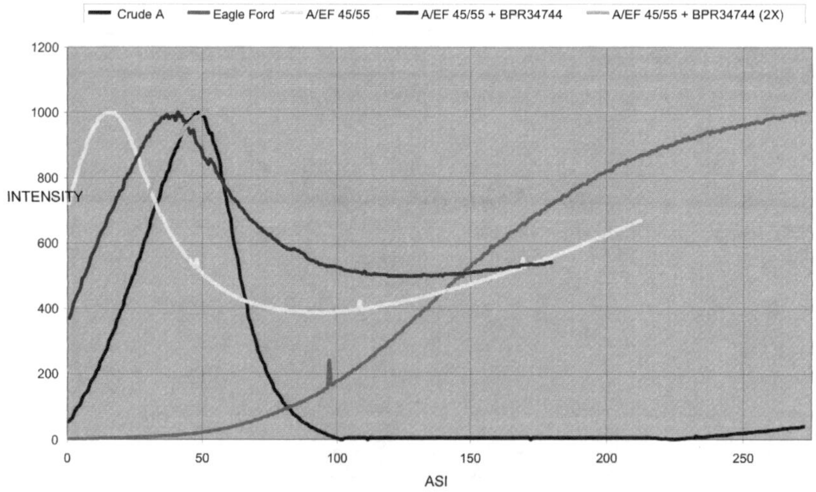

14 AIChE Spring Meeting, New Orleans, LA

BAKER HUGHES

Tower Tray Fouling - Phosphorus

- Seen by refineries running Bakken, Eagle Ford, and WTI
- Operationally manifested as increasing tower dP
  - Typically in the kero/jet zone
- Tower inspection shows deposits on trays in kero/jet zone
  - Some deposition may also show a few tray levels above and/or below kero/jet zone
  - Deposit scale typically tightly bonded to tray surface restricting flow and valve function

Above kero/jet zone

In kero/jet zone

Below kero/jet zone

14AIChE® Spring Meeting, New Orleans, LA

BAKER HUGHES

Shale Oil Impact – MEA Contamination

Corrosion and Fouling

- Shale oils transported by truck, rail, barge.
- Many shipments are treated with H_2S scavengers:

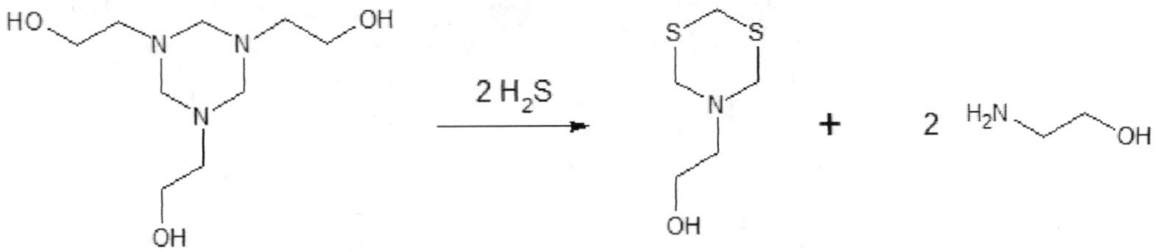

- Potential Impacts:
  - H_2S scavengers leave byproduct amine in crude oil (MEA, MA).
  - Tramp amines distill to OH system, create amine-HCl salts:
  - Corrosive to system metallurgy.

14AIChE® Spring Meeting, New Orleans, LA

BAKER HUGHES

Impact of Tramp MEA on OH Water

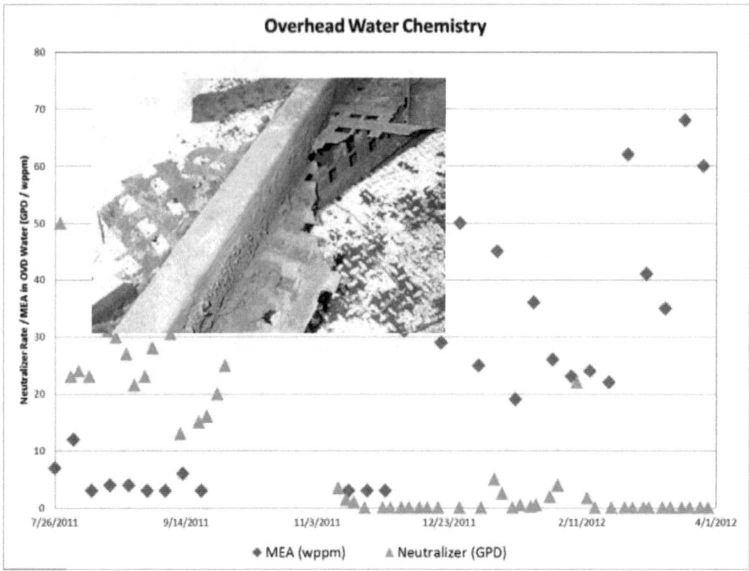

14 AIChE Spring Meeting, New Orleans, LA

BAKER HUGHES

Salt Risk w/MEA Concentration

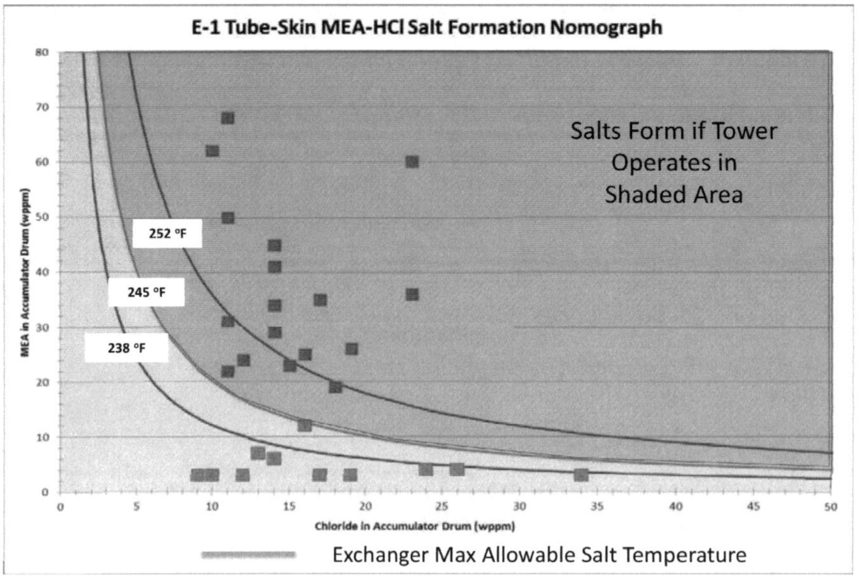

14 AIChE Spring Meeting, New Orleans, LA

BAKER HUGHES

Shale Oil – Refinery Process Concerns and Monitoring

BAKER HUGHES

CHARACTERISTIC OF SHALE OIL	CONCERN IN REFINERY	PROCESS DATA	SAMPLES AND ANALYSES: Samples	SAMPLES AND ANALYSES: Analysis	OUTPUT DATA	FREQUENCY	BHI PROGRAMS
Crystalline Wax	Tank bottom sludge	• Tank Sludge Depth	• Tank bottom	• Visual sludge • GC analysis	• Tank Sludge Depth • Crystal Wax Presence	• Monthly	• XERIC™ Pretreatment
	Cold Preheat Fouling	• Desalter Temperature			• Temp vs. Time	• Update 2x/mo.	• LIFESPAN™ Wax Dispersants
	Desalter emulsions	• Emulsion layer growth	• Triline emulsions	• Visual wax • GC analysis	• Crystal Wax Presence	• After Shale Oil begins • When emulsion grows	• XERIC Demulsifiers
Asphaltene Stability in Blends	Desalter emulsions	• Emulsion layer growth	• Triline emulsions	• Precipitated Asphaltenes	• Precipitated Asphaltenes	• After Shale Oil beings • When emulsion grows	• XERIC Asphaltene Stabilizers • Field ASIT services™
	ColdPreheat Fouling	• Desalter temperature	• Crude blend	• ASIT	• Temp vs. Time • ASIT #	• Pre-Shale Oil • Update 2x/mo.	• LIFESPAN Antifoulants • Field ASIT services
	Hot Preheat Fouling	• Furnace inlet temp • Furnace tube temps • Furnace fuel usage • Preheat DP	• Crude blend	• ASIT	• Temp vs. Time • Fuel use vs. Time • ASIT #	• Pre-Shale Oil • Update 2x/mo.	• LIFESPAN Antifoulants • MILESTONE™ Furnace Fouling Control • Field ASIT services
Paraffinic	Furnace Coking	• Furnace tube temps • Furnace fuel usage	• Crude blend	• HLPS	• Coking tendency	• Pre-Shale Oil • W/Shale Oil	• MILESTONE Furnace Fouling Control
Tramp MEA	Tower/OH corrosion (Early Warning)	• OH Neutralizer rate • EXCAL usage rate	• OH water • Desalter brine	• OH pH • Brine pH	• Early Sign of Amine • Early sign of Amine	• Pre-Shale Oil • W/Shale Oil	• TOPGUARD™ Overhead Corrosion Control • EXCALIBUR™ Contaminant Removal Technology
	Tower/OH corrosion (Quantitative)	• Inputs for CRM	• OH water • Desalter ww • Desalter brine	• IC Amines	• Confirm MEA in Crude • OH MEA ppm • Salt Risk Tower • Salt Risk OH System	• Pre-Shale Oil • W/Shale Oil • Update 2x/mo.	• TOPGUARD Corrosion Control • EXCALIBUR Technology • TOPGUARD Corrosion Risk Monitor (CRM)
Phosphorus	Tower P Fouling	• Tower DP			• Tower DP Increase	• 2x/mo.	• LIFESPAN Antifoulants, Dispersants
			• Crude	• ICP for P	• P in Crude	• 2x/mo.	
			• Kerosene	• Filt Solids • XRF Filter P • ICP for P	• Incr. Kero Solids • P in Solids • P in Kerosene	• 2x/mo.	
			• Kerosene • Diesel	• ASTM D86	• Incr. in Diesel IPB – Kero EP overlap	• 2x/mo.	

EXCALIBUR, FIELD ASIT SERVICES, LIFESPAN, MILESTONE, TOPGUARD and XERIC are trademarks of Baker Hughes Incorporated.

Development of a Fundamental Wax Deposition Model for Water-in-Oil Dispersed Flow

Mark Zheng and H. Scott Fogler

Wax Deposition in Off-shore Productions

2

- A challenge for flow assurance
  - Wax deposit at T_{wall} < WAT
- Significant costs related to remediation practice
  - Pigging
- Deposition modeling
  - Fundamental understanding
  - Enable effective remediation

www.deepstar.org

Niesen, E&P Magazine 2002

wax deposition in subsea pipelines

cost as a function of pigging frequency

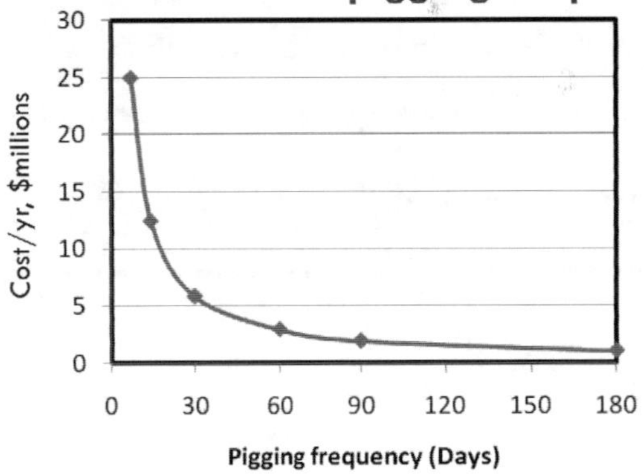

The Michigan Wax Predictor (MWP) – the Most Advanced Wax Deposition Modeling Tool

3

- Developed through
  - 4 Ph.D. thesis at the University of Michigan over 15 years
- Verified by
  - flow-loop experiments from different facilities around the world
- The **only model** that solves the fundamental transport equations
- Predicts **consistent trends** of deposition rates with varying operating conditions
- Contains **no tuning parameter** for the upper and lower limits
- Contains **one tuning parameter** to account for wax precipitation in the bulk

Deposition Rate Predictions by The Michigan Wax Predictor (MWP)

4

Garden Banks Crude Oil

South Pelto Crude Oil

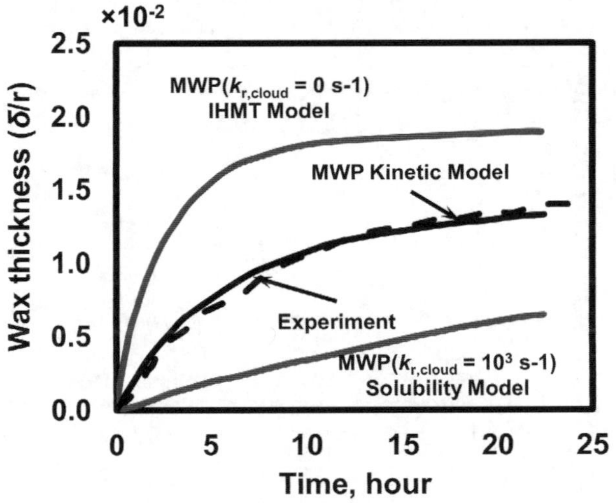

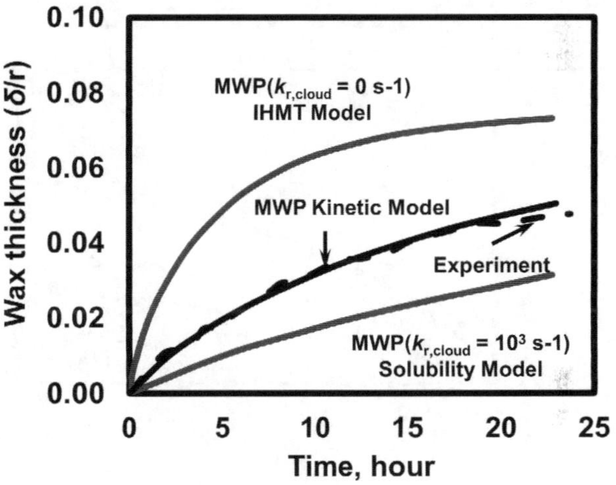

- Without tuning, MWP predicts upper and lower limits that bound the actual deposition rate.
- Predictions can match experimental deposition rate by tuning the bulk precipitation kinetic constant.

Huang et al. (2011)

More Deposition Rate Predictions by The Michigan Wax Predictor (MWP)

5

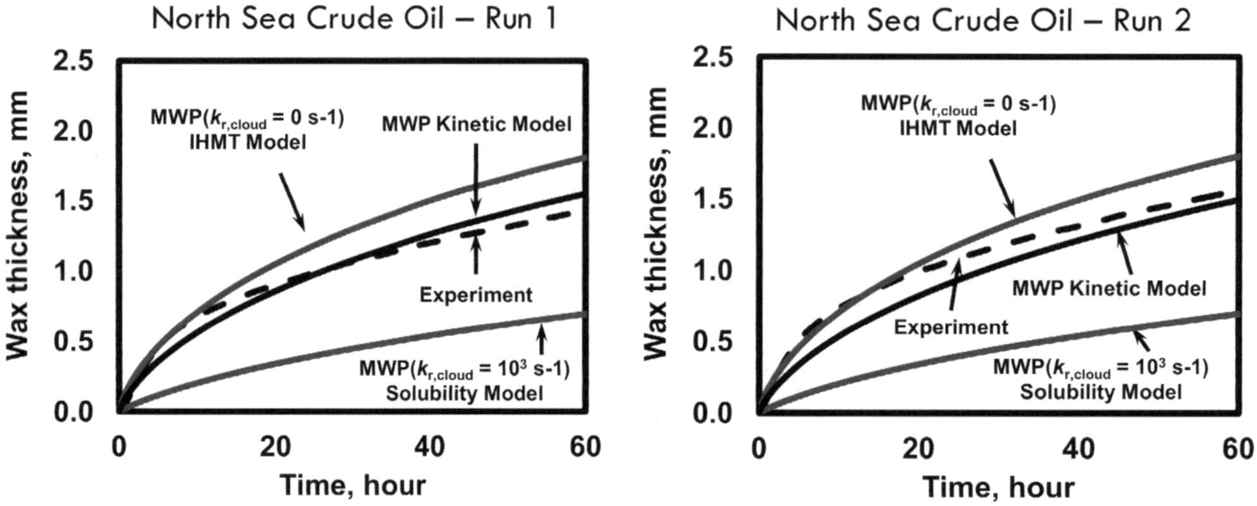

- Predicts consistent deposition rate with >20 more deposition experiments.

Huang et al. (2011)

Current Status of Wax Deposition Modeling in Water-in-Oil Dispersed Flow

6

- In order to model effects of water on the heat and mass transfer characteristics
  - Heat transfer: pseudo-single phase approach
  - Mass transfer: **no fundamental modeling approaches due to difficulties (see next slide)**
- Goal: based on MWP, build a fundamental wax deposition model for dispersed flow.

The Effect of Water Droplets on Mass Transfer: Partially Block the Diffusion Path

7

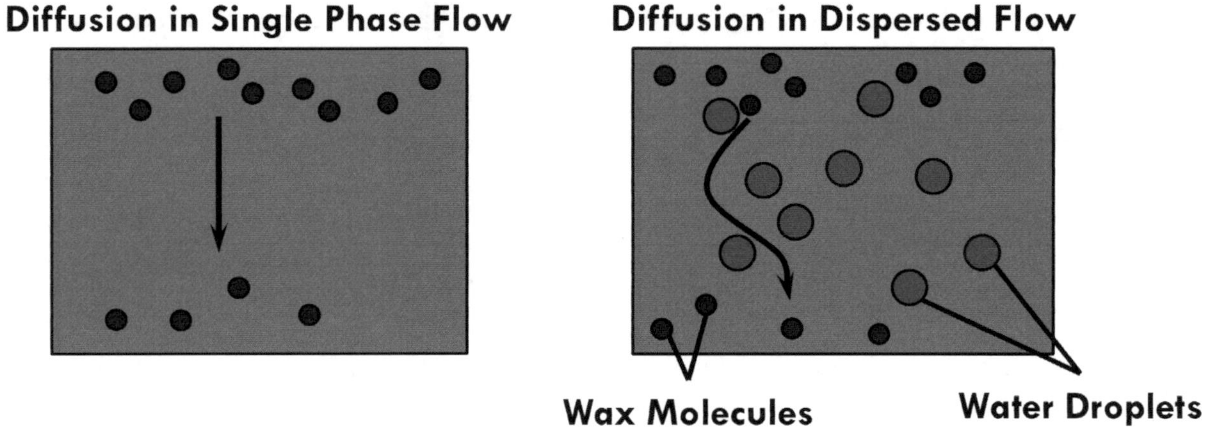

- Difficult to quantify the effect of water droplets on wax diffusion
- No established modeling strategies

Current Strategies for Modeling Diffusion in Water-in-Oil Dispersion

8

- Strategy 1: Use emulsion viscosity for diffusivity calculation
  - Does not represent the physics: diffusion occurs in the continuous phase
- Strategy 2: Use continuous phase diffusivity
  - Does not capture the effect of water droplets on diffusion.
- Neither approach is fundamentally correct

energy&fuels Review
pubs.acs.org/EF

Review of Paraffin Deposition Research under Multiphase Flow Conditions

Cem Sarica* and Ekarit Panacharoensawad

It is still not clear if the mixture viscosity or the oil viscosity should be used for the wax diffusivity calculation in the Hayduk–Minhas equation[61] because this equation is not

Sarica et al. (2012)

Outline of the Project

9

- Investigation of diffusivity reduction due to water droplets
  - Verify
  - Quantify
  - Model

- Based on water-in-oil model emulsion

Preparation of Model Water-in-Oil Emulsion

10

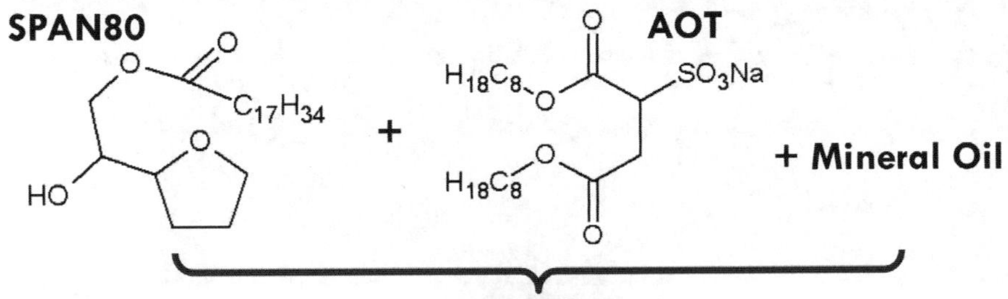

Surfactant:
5wt.% of total emulsion
(SPAN80: AOT = 9:1)

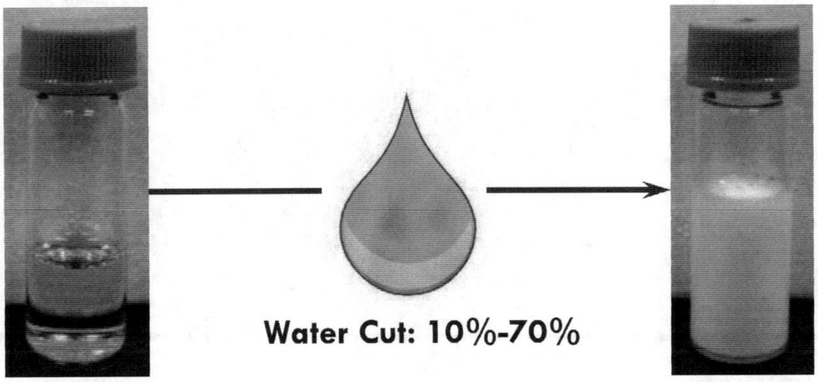

Delgado-Linares et al. (2013)

"Bulk Stability" of Emulsion

11

Time t = 0 hr

Time t = 48 hr

- Emulsion samples (Water cut = 10-70%) are stable over 48 hours.

Microstructure Stability – 1: Mean Droplet Size Constant Over 48 Hours

12

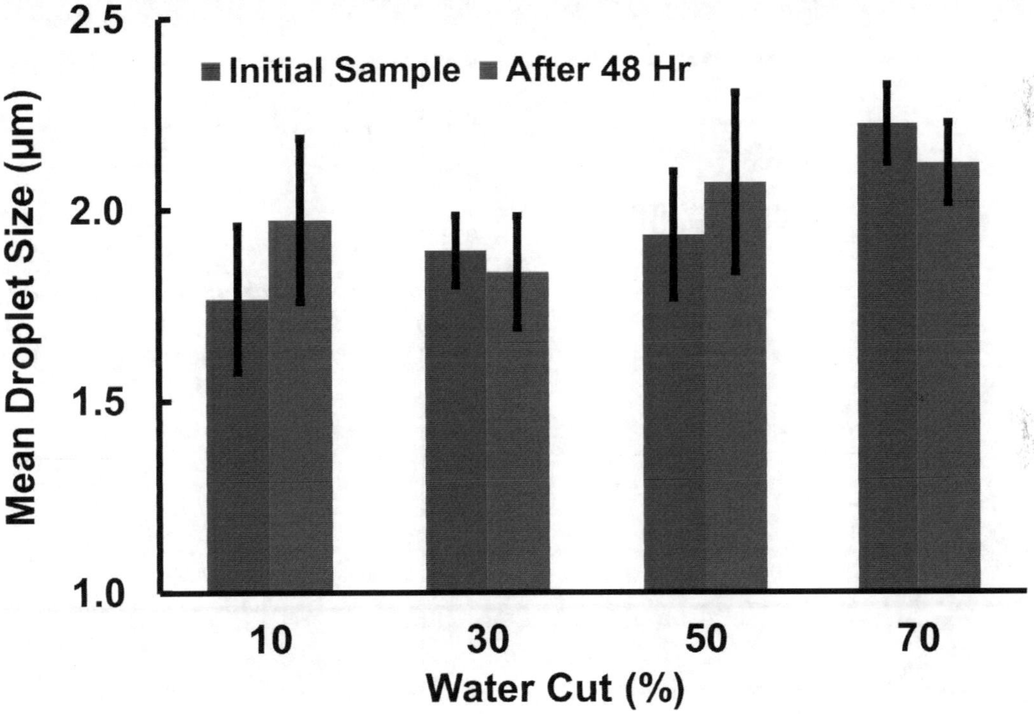

Microstructure Stability – 2: Droplet Size Distribution Constant Over 48 Hours

13

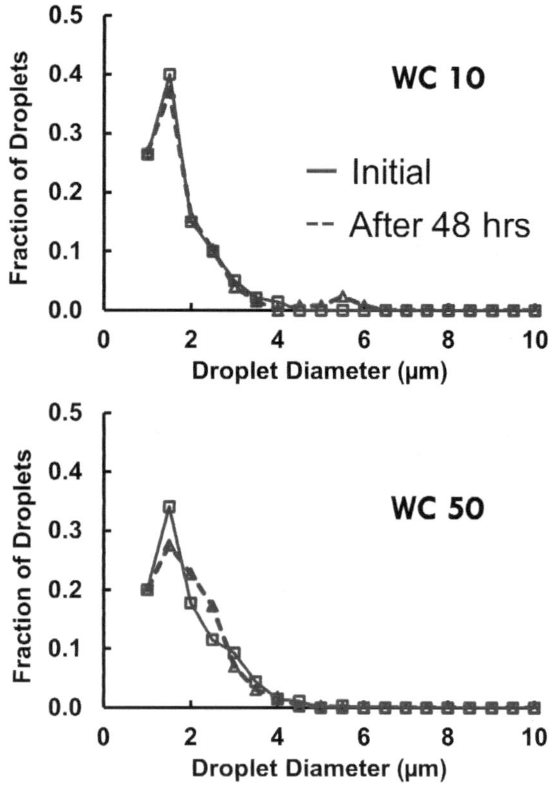

Summary

14

- Prepared emulsion with water cut ranging from 10 to 70% with 5wt.% surfactant

- Emulsions are stable for at least 48 hours
  - Macroscopically
  - Microscopically

- Can characterize the diffusivity of solute in emulsion
  - Study the effect of water drops on solute diffusion
  - Have a 48-hour window

The Idea of Using NMR to Characterize Diffusivity in Emulsion

15

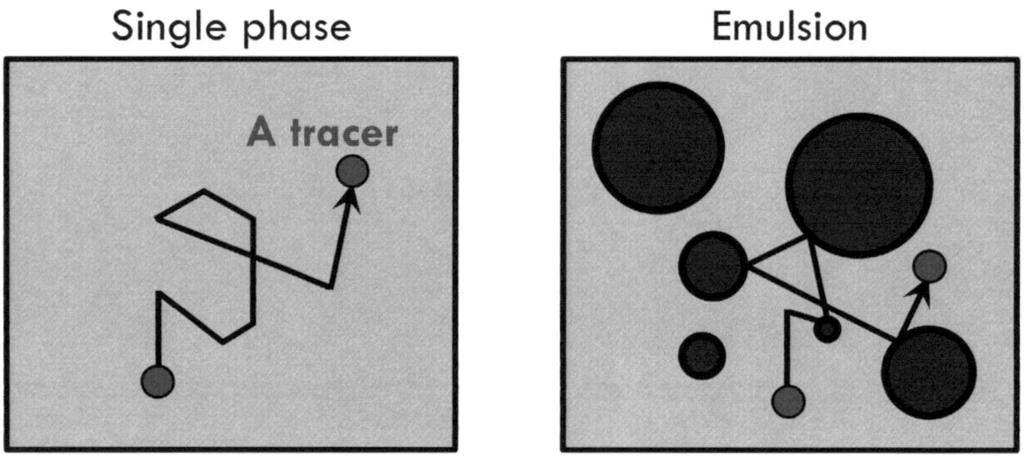

D_{single} D

- Expected outcome: $D < D_{single}$
- Quantify the diffusivity reduction by calculating D/D_{single}

Identification of Tracer in Emulsion Mixture Using Its Chemical Shift

16

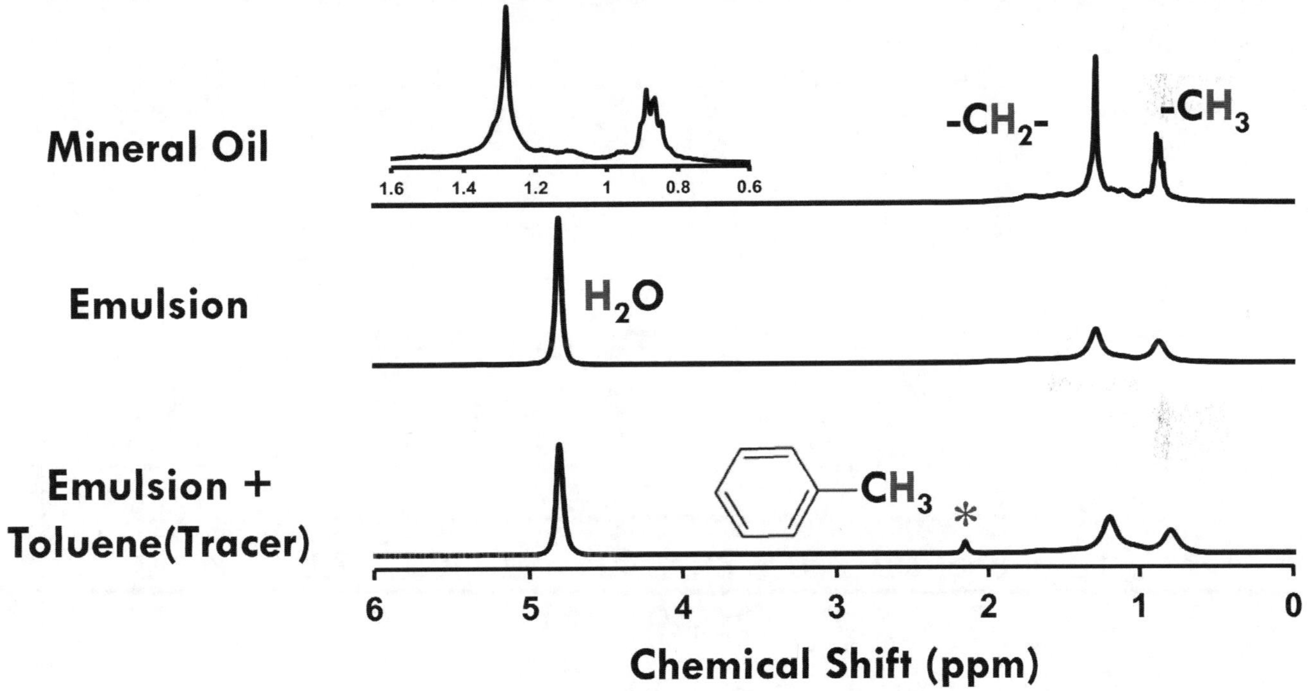

Characterization of Toluene Diffusivity in Emulsion

17

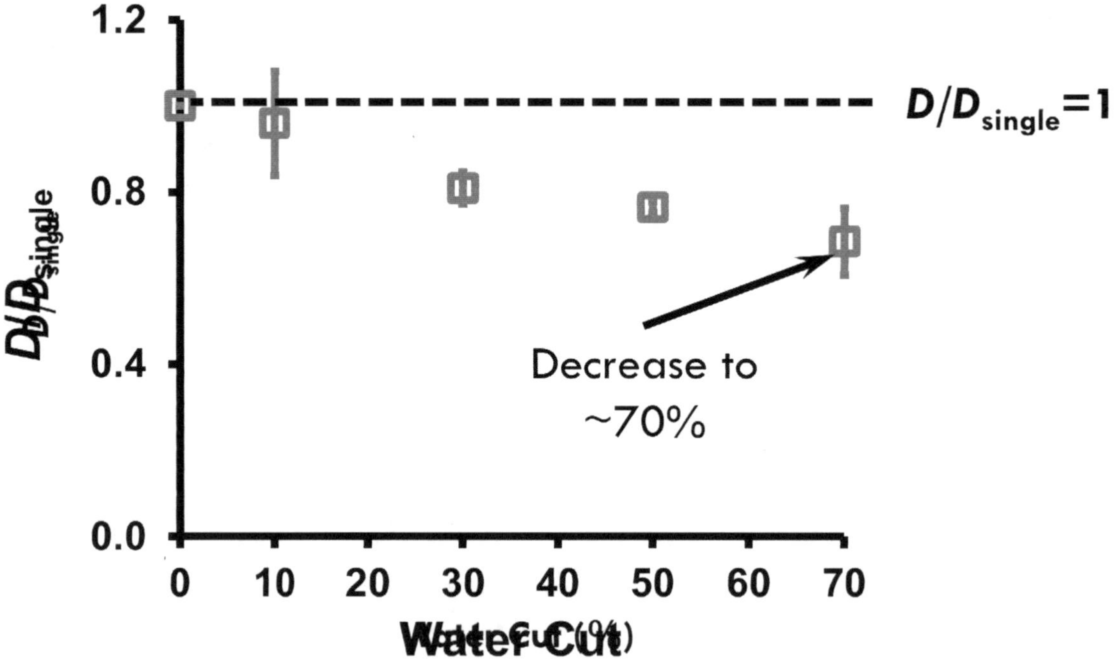

- At WC=70, diffusivity decreases to ~70% single phase diffusivity.

The Species Dependency of Diffusivity Reduce in Emulsion

18

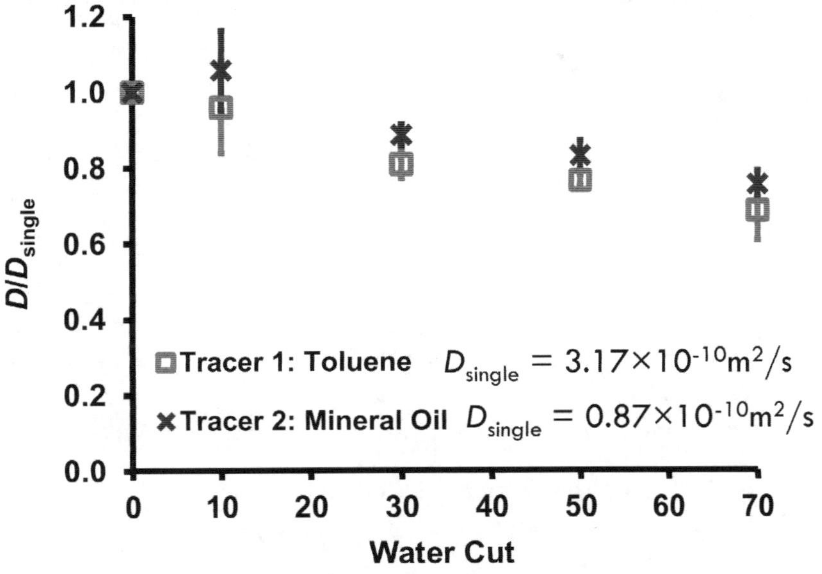

- The reduction in diffusivity seems to be "species-independent".
- Is it possible to measure wax diffusivity in emulsion?

Characterization of "Wax" Diffusivity in Emulsion – Part I. Difficulties

19

- Difficulties:

Wax, e.g. $C_{20}H_{42}$:

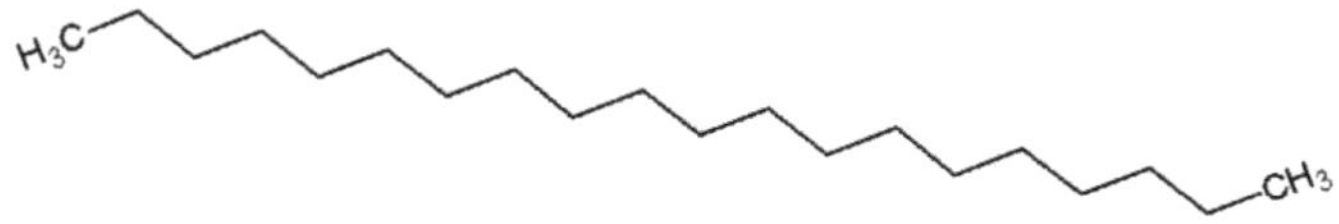

Mineral Oil: -CH, $-CH_2$, $-CH_3$

- Chemical structures of wax and mineral oil are similar
- Cannot distinguish between wax ^{1}H-NMR signal from the "mineral oil ^{1}H-NMR background"

Characterization of "Wax" Diffusivity in Emulsion – Part II. Solutions to Difficulties

20

- Use 1-fluorine-hexadecane to represent wax (F-Wax)

Regular wax:
cannot be detected in mineral oil medium

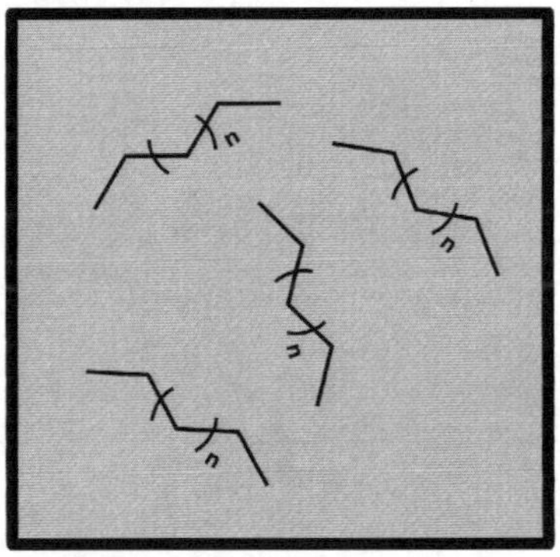

F-Wax:
can be detected in mineral oil medium

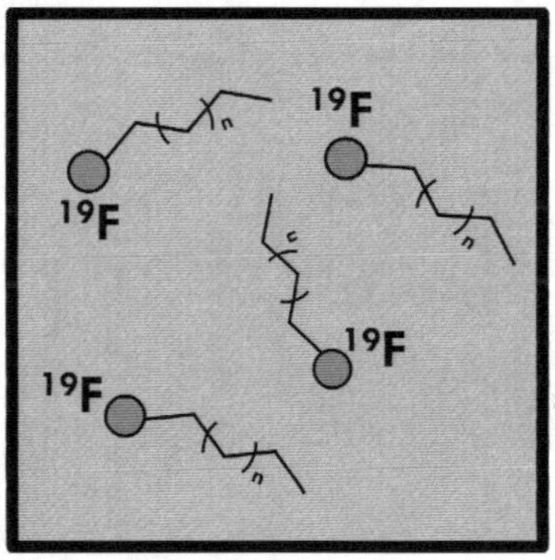

- "^{19}F tagged"
- Can track the diffusion of "F-Wax" by ^{19}F-NMR

Characterization of “Wax” Diffusivity in Emulsion – Part III. Results: Identification of F-Wax in ^{19}FNMR

21

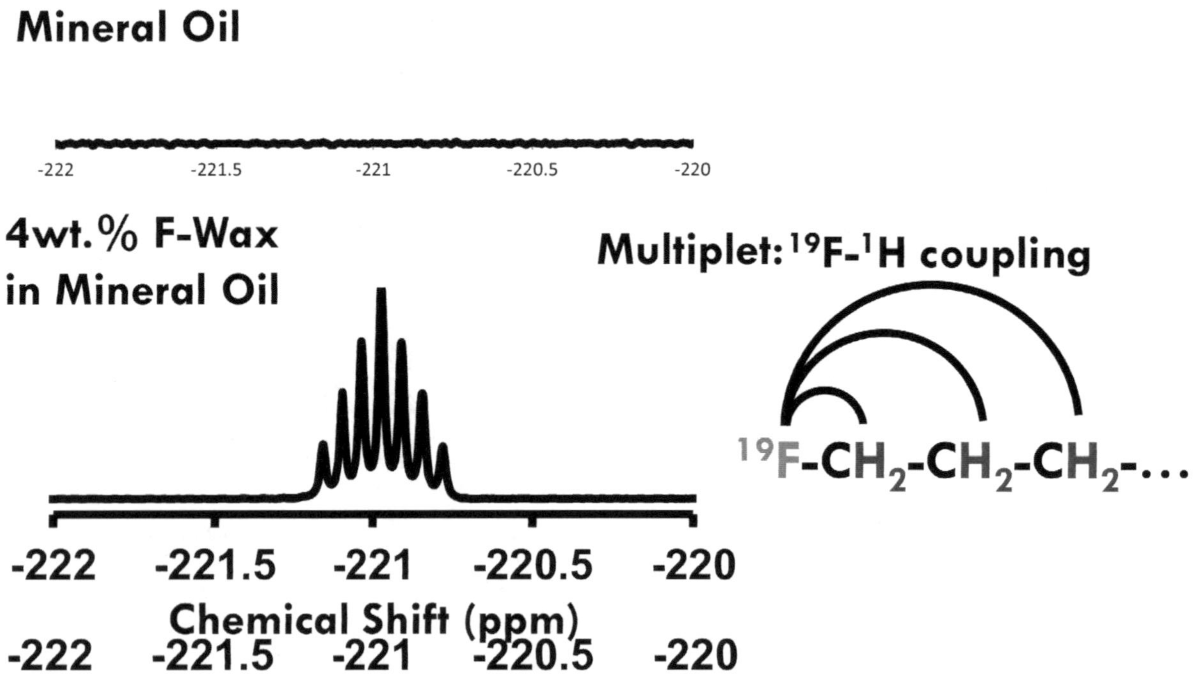

Characterization of "Wax" Diffusivity in Emulsion – Part III. Results

22

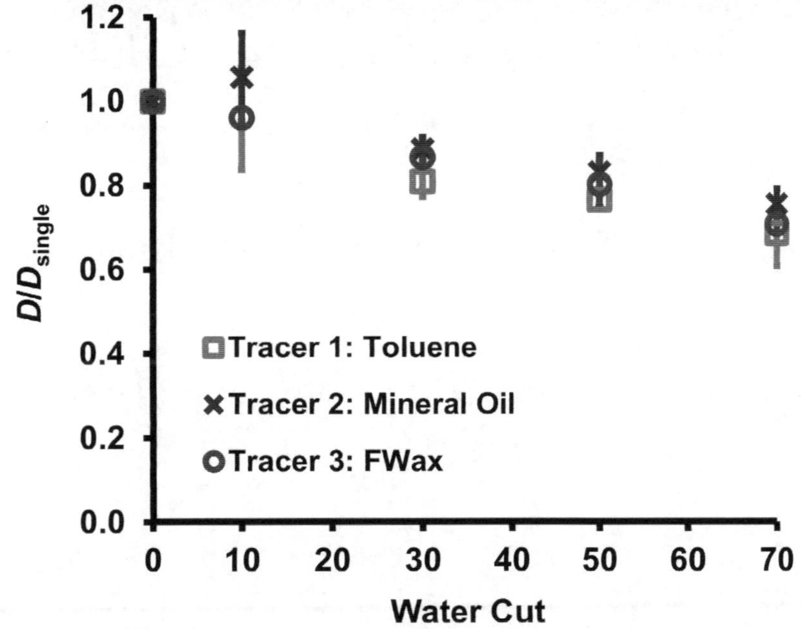

- Diffusivity reduction of a "wax-like" molecule and tracer molecules are similar.

Summary

23

- The reduction in diffusivity (D/D_{single}) due to droplets is species-independent.

- D decreases to ~70% D_{single} in WC70 emulsion

- Further investigation to model this diffusivity reduction

Strategy 1 for Diffusivity Calculation in Dispersed Flow

- Use emulsion viscosity in single phase diffusivity correlations

Hayduk-Minhas:

$$D \sim \frac{1}{\mu^{0.791-10.2/V_A}} \sim \frac{1}{\mu^{0.77}} \quad (V_A \sim 400 \text{ L/mol for wax})$$

Wilke-Chang:

$$D \sim \frac{1}{\mu}$$

Investigate the Effect of Using Emulsion Viscosity for Diffusivity Calculation

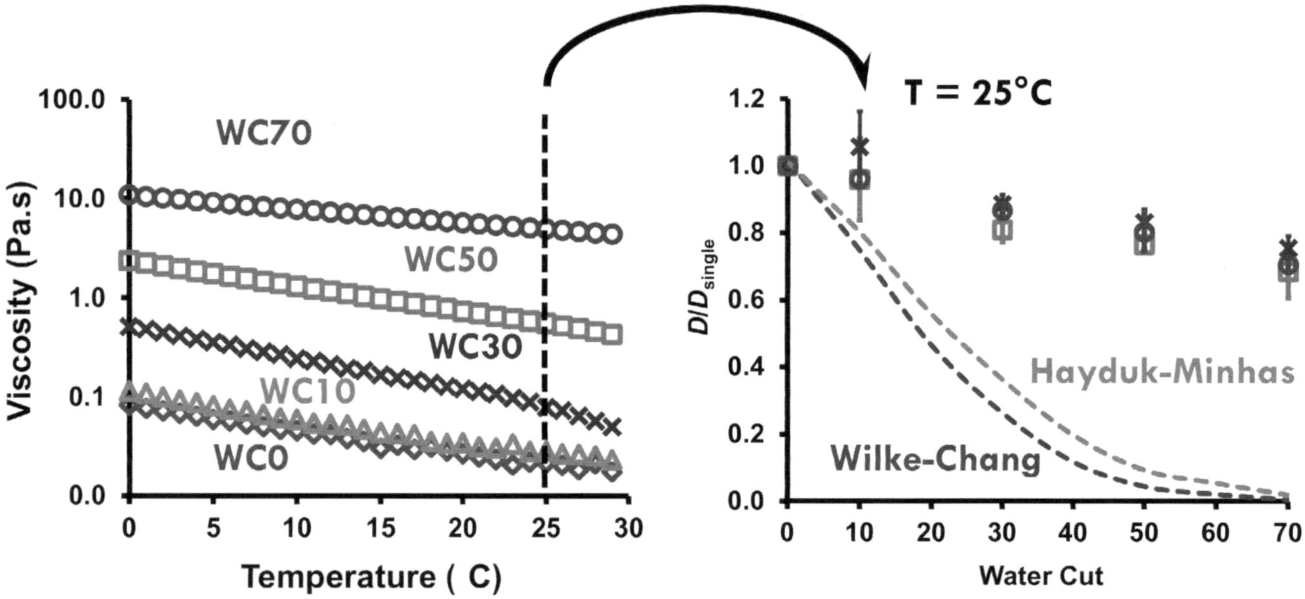

- Using emulsion viscosity to calculate diffusivity leads to 2 orders of magnitudes under-prediction.

Improved Strategy for Diffusivity Calculation in Dispersed Flow

26

Aris and Cussler (diffusion across membranes):

$$\frac{D}{D_{\text{single}}} = \frac{1}{1 + \frac{\alpha^2\Phi^2}{1-\Phi} + \frac{\alpha\Phi}{\sigma} + \frac{4\alpha\Phi}{\pi(1-\Phi)}\ln\frac{\pi\alpha^2\Phi}{\sigma(1-\Phi)}} \qquad \frac{D}{D_{\text{single}}} = \frac{1}{1 + \frac{\alpha^2\Phi^2}{1-\Phi}}$$

Jönsson (diffusion around spherical barriers):

$$\frac{D}{D_{\text{single}}} = \frac{2}{2+\Phi}$$

Aris et al. (1985), Cussler et al. (1988), Jönsson et al.(1986)

Predictions by Aris' Equation and Cussler's Equation

27

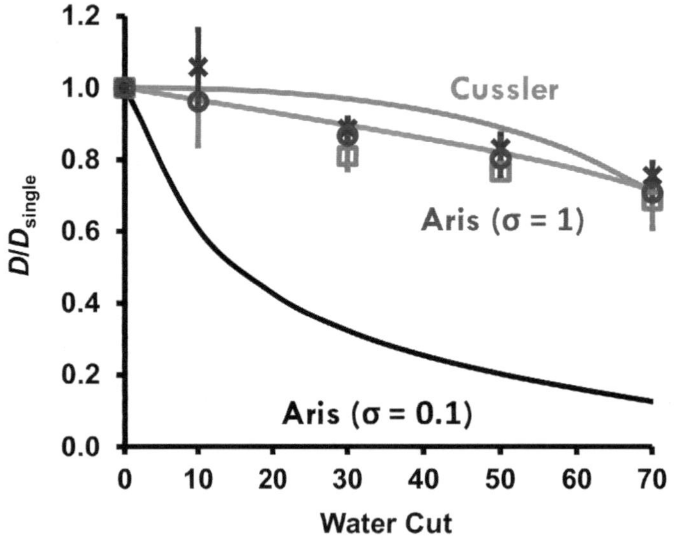

- Predict the correct order of magnitude
- Characterization of the spacing between droplets is difficult

Predictions by Jönsson's Equation

28

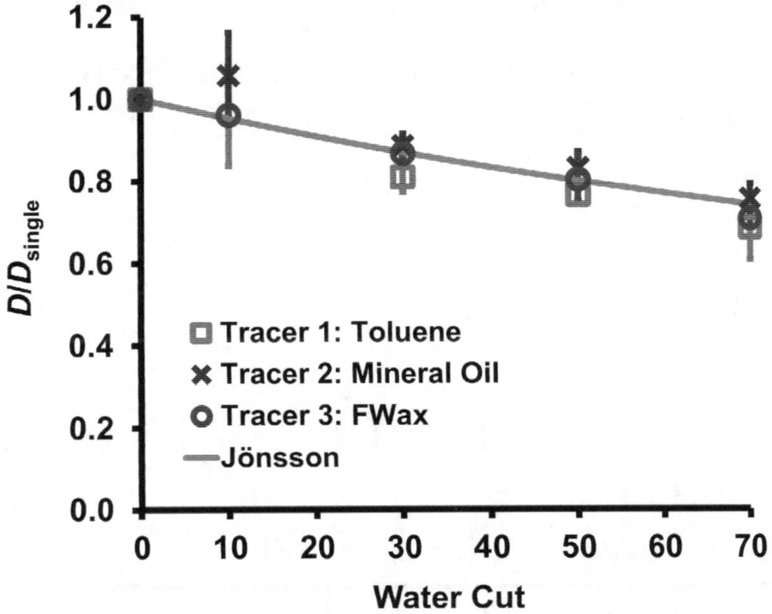

- Estimation by Jönsson's equation is in good agreement with experimental observations.

Summary

29

- Using emulsion viscosity to calculate the diffusivity causes two orders of magnitudes deviation in D/D_{single}.

- The effective diffusivity can be estimated based on the continuous diffusivity.

- Development and validation of the model in progress
  - Incorporation of dispersed phase diffusivity calculation into the Michigan Wax Predictor (MWP)
  - Validate by flow loop experiments

Acknowledgements

30

TOTAL

ConocoPhillips

Chevron

NALCO Champion

An Ecolab Company

msikenny

bp

multi-chem

A HALLIBURTON SERVICE

HALLIBURTON

Statoil

352143 Thermal Cracking Partial Upgrading Product Unstable and Incompatible

Abstract:

Bitumen and extra heavy oil containing asphaltenes are usually blended with condensate, light conventional oil, and, or fully upgraded, hydrotreated synthetic crude oil (SCO) for pipeline transportation and sale to the refinery market. Research and practice keep the ratio of light oil, condensate or SCO and the bitumen to the portions that will not cause the precipitation and deposition of asphaltenes. Some dispersants, flow improvers or additives are used in the bitumen blends. With blending as the most standard, common method of improving viscosity and density, there is little risk to this production transportation method.

Recently heated rail cars are used to transport bitumen without diluent. As the ratio of asphaltenes to other components such as resin does not change, there is little risk of precipitation of asphaltenes in this production transportation method.

During the period of high light crude price, low heavy crude price, high heavy-light crude differential, high diluent and high natural gas cost, partial upgrading schemes were proposed for in-situ down hole and at the wellhead. Typically, the partial upgrader proposals have not provided any flow assurance data that show asphaltenes do not precipitate or deposit in the process heaters, reactors, pipes, vessels, exchangers, downstream, or in storage.

Over twenty in-situ down hole methods are in various stages of pilot and testing techniques that might result in some partial upgrading. Over forty field located partial upgrading schemes have been proposed, none have been build or otherwise commercially proven. Down hole or above ground, the goal is to produce a lower viscosity material at the well-site, upstream, or midstream, that would require less diluent or natural gas. For example Fluid Catalytic Cracking, FCC, analogues UOP-CCU, Catalytic Crude Upgrading, Viscositor, Ivanhoe Energy IE-HTL, ETX-IYQ, NZ cracking use FCC configurations without hydrotreating the reaction product. For example, HTL uses high temperature short contact time to thermally crack bitumen vacuum bottoms.

Asphaltene deposition and diene gum formation are major flow assurance risks in partial upgrading thermal processing like HTL with high temperature short contact time reactions in an FCC analogue circulating sand.

The above ground field located partial upgrading methods have no commercial installations to validate performance. Despite claims of tens and hundreds of runs and commercial readiness, there is very little pilot plant or demonstration unit data on the product properties of the various methods; there is virtually no flow assurance data on the stability or compatibility of the various methods proposed for partial upgrading.

One thermal cracking partial upgrading process claims to have stable product, but is unable provide any data to demonstrate stability using any of the sixty possible test methods that have been used to show stability of fuel oils, crude, thermally and, or catalytically cracked products. On the other hand, problems associated with thermal reaction products are well known from the earliest days of refining with Shukhov, Burton and Dubbs process units. Unstable thermal reaction products at higher temperatures have been recognized and addressed throughout the long history of delayed, fluid and flexi coking, visbreaking, thermal and catalytic cracking.

High temperature thermal cracked reactions yield unstable products and with bitumen resid as feed most yield asphaltenes that are self-incompatible, which is completely different than hydrotreated, stable, and compatible SCO. Rather than overstate, confuse, or misrepresent the thermally degraded quality, partial upgraded material should be named differently from SCO, for example, Partially Upgraded Thermal Cracked Unstable Product would have acronym PUTCUP. PUTCUP with high olefins and dienes, disallowed by some pipelines, could be shipped like raw bitumen in rail cars, or if olefins allowed could be shipped like dilbit in pipelines or rail cars after blending with diluent to gravity or viscosity specification. PUTCUP includes heavy gum forming dienes in naphtha, jet and distillate fractions, self-incompatible asphaltenes and inorganic feed and produced solids in all fractions.

Patents and other open literature sources show short contact, high temperature reaction process products, like FCC, fluidcoker, HTL and pyrolysis naphtha and distillates, are less stable than lower temperature longer contact time products like visbreaker and delayed coker naphtha and distillates. These facts are in contrast to unsupported marketing claims, like HTL, of stable product due to shorter time. Simple chemistry and physics, it's about temperature, the higher the reaction temperature the more unstable product is formed, as shown by reaction product gases that have higher dienes including 1,3-butadiene. The higher the short contact reaction temperature, the more 1,3-butadiene in product gas with more liquid dienes throughout the reactor product, and the less stable the product. Thermal cracking processes with higher mix zone temperatures like HTL produce higher 1,3-butadiene, and corresponding higher C5-C10 dienes leading to high gum formation than process with lower reaction mix zone temperatures.

Processes, like HTL, with high temperature short contact time of fractions of seconds result in immediate reactor and downstream plugging with unstable self-incompatible asphaltenes reaction product precipitating and depositing with inorganic feed and produced solids and downstream heavy gum formation and deposits caused by dienes in naphtha, jet and distillate fractions. Thermal cracked product gum and residual asphaltenes deposit and plug reactors, transfer lines, quench towers, fractionation towers, downstream filters, catalyst guard beds, hydrotreaters, pumps, pipes and tanks. Self-incompatible PUTCUP is worse than just incompatible with other liquids as PUTCUP has precipitated, condensed and flocculated asphaltenes, semi-solids and solids that are out of solution, toluene insoluble, and cannot be resuspended or dissolved by any dispersant, flow improver, or other additive available.

No partial upgraders have been built to prove reliable, long term flow assurance of a year or more continuous operation, rather than a few hours of pilot or demonstration unit run before terminated due to plugging, etc.. Proof of reliable operation with industry standard on-stream factors over 95% uptime is necessary to reduce high risk of a high capital cost, ~ \$550 million, investment in an unproven process. The ~\$550 million for a 20,000 bpd partial HTL upgrader is like other partial upgrader estimates that typically don't include hydrotreating in order to show a significantly lower capital cost compared to full upgraders. However, partial upgrading reactor product is instantaneously self-incompatible with precipitated or condensed asphaltenes depositing and plugging in and on reactor walls, transfer lines, quench towers, pipes, and vessels downstream, and is inherently unstable due to dienes that form gums downstream of the reactor. PUTCUP fractions are thermally degraded compared to crude or bitumen. PUTCUP properties are similar to fluid coker product fractions.

Reduced flow assurance risk solutions to these partial upgrading challenges include the common practice of shipping bitumen in heated rail cars, or blending bitumen with diluent without partial upgrading.

Modeling Wax Deposition with Advanced Heat and Transfer Algorithms

(Jason) Zhenyu Huang, PhD
AIChE 2014 Spring Meeting

Assured Flow Solutions LLC

Introduction

Assured Flow Solutions LLC

- Wax deposition challenges
  - Wax deposit as Toil < WAT
  - A problem of 10^7
- Significant remediation costs
  - Pigging
  - Thermal insulation
  - Inhibitors
- Deposition modeling
  - Fundamental understanding
  - Enable effective remediation

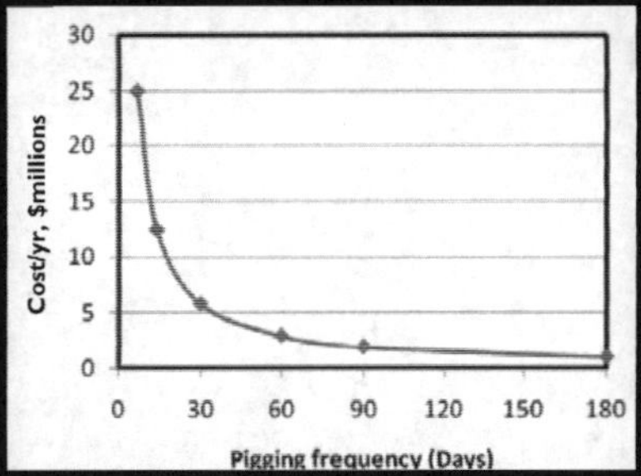

www.deepstar.org
Niesen, E&P Magazine 2002

Modeling Wax Deposition

Assured Flow Solutions LLC

Thermodynamics

Transport Phenomena

- How much does wax precipitate at a given temperature?
- How much does wax deposit at a particular location?

Wax Thermodynamics

Assured Flow Solutions LLC

- Waxy Components
  - n-C17+
- Wax Precipitation
  - WAT
  - Wax content
- Important Info to Further Investigate Wax Deposition Severity

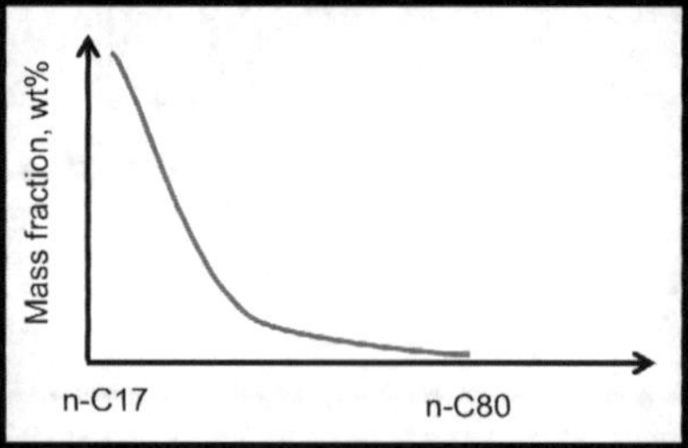

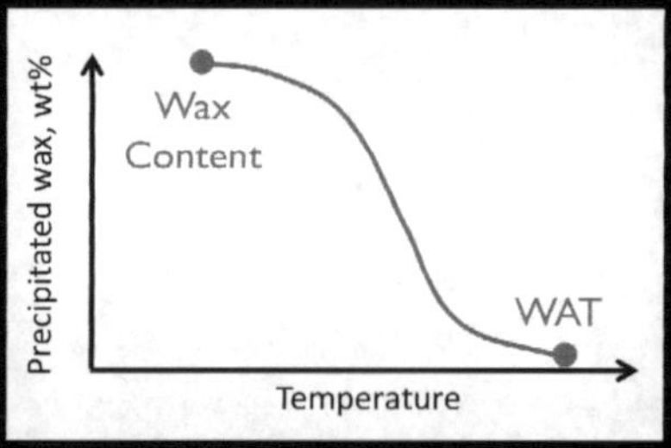

Understand Wax Thermodynamics

Assured Flow Solutions LLC

- Theoretical Modeling
  - Solid-liquid equilibrium of n-paraffins

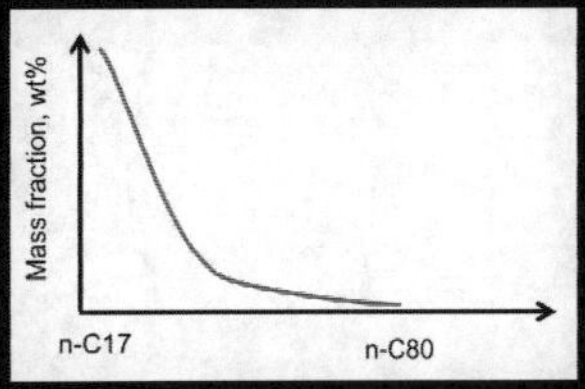

Equilibrium Types		PVTSim	GUTS	Multiflash
Vapor-Liquid Equilibrium	Vapor	Non-ideal, SRK Equation, (Won 1986)		
	Liquid			
Solid-Liquid Equilibrium	Solid	Single ideal solution	Multiple ideal solid solutions	Single or multiple non-ideal solid solution
	Liquid	Non-ideal (SRK)	Ideal solution	Non-ideal, Flory free volume theory (Coutinho, 1996)

Pedersen, K.S. SPE Production & Facilities, February 1995:46-49
D.D. Erickson, V.G. Niesen, and T.S. Brown, 1993, SPE 26604.
Coutinho, J.A.P., Stenby, E.H., Ind. Eng. Chem. Res. 1996;35:918

Understand Wax Thermodynamics

Assured Flow Solutions LLC

- Experimental Investigation
  - Cross-Polarized Microscopy (CPM)
    - Mostly for WAT
  - Differential Scanning Calorimetry (DSC)
    - WAT and precipitation curve
- Provide Wax Thermodynamics at stock tank conditions
- Important for Thermodynamic Model Tuning

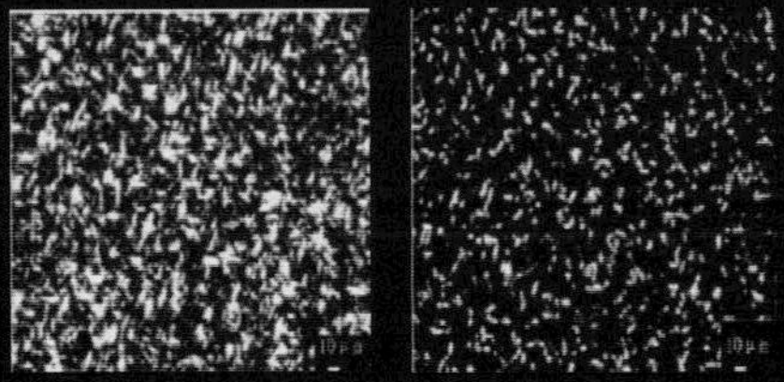

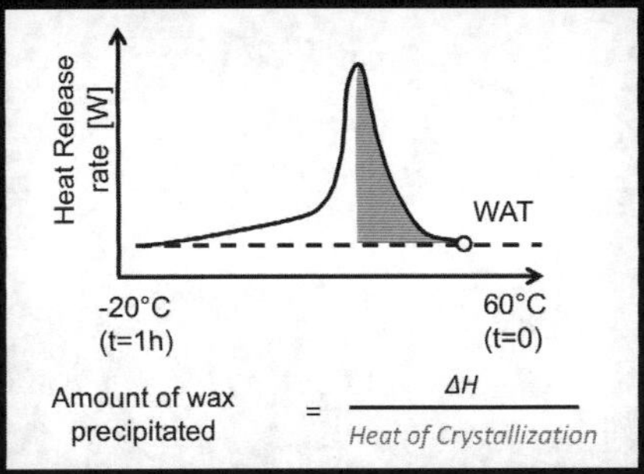

Pedersen, K.S. SPE Production & Facilities, February 1995:46-49
D.D. Erickson, V.G. Niesen, and T.S. Brown, 1993, SPE 26604.
Coutinho, J.A.P., Stenby, E.H., Ind. Eng. Chem. Res. 1996;35:918

Wax Deposition – Transport Phenomena

Assured Flow Solutions LLC

- Heat Transfer
  - Temperature distribution of the pipe
- Mass Transfer
  - What happens to the waxy components in the pipe?
- Fluid Dynamics
  - Gas/Water/Liquid flow

Studies on Wax Deposition Modeling

Assured Flow Solutions LLC

Earliest studies (1980s) (Understand Wax Deposition Mechanism)	Newberry, 1984 Majeed, 1990 Hsu, 1995
Studies around 2000 (Develop industrialized and academic models)	Rygg, 1998 (OLGA Wax) Matzain 2001 (OLGA Wax) Lindeloff and Krejberg 2002 (PVTSim DepoWax)
Recent academic studies (Understanding the transport phenomena in wax deposition)	Singh, 2001 Vekatesan 2004 Lee, 2008 Huang, 2010, 2011 Sheng, 2013 } The Michigan Wax Predictor

A gap seems to exist between industrial models and academic developments

Comparison on Wax Models

Assured Flow Solutions LLC

- Difficult - Lack of Good Data
  - Different apparatus were used
  - Need well-designed experiments covering a series of factors
    - (Oil, flow rate, temperature)
- Recent Study by Hoffmann et al.
  - 3 Years of experimental development
  - Indoor 2", 5.5m flow-loop
  - 50+ wax deposition experiments
  - Allow comprehensive model comparisons

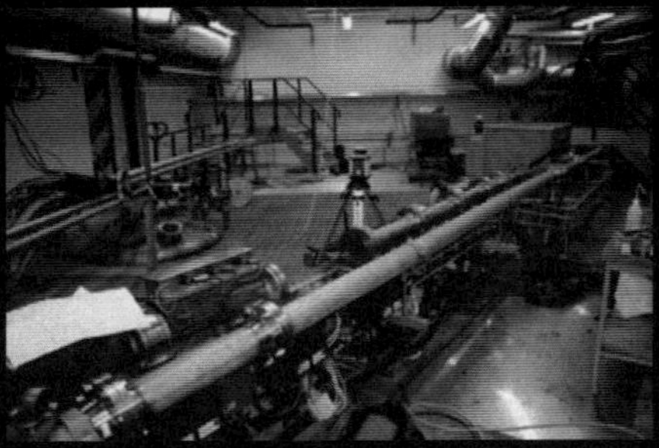

Evaluating Wax Deposition Models

Assured Flow Solutions LLC

- Compare Industrial and Academic Wax Models
  - OLGA Wax
  - MWP
- Oil studied
  - Waxy condensate from North Sea (30C WAT)

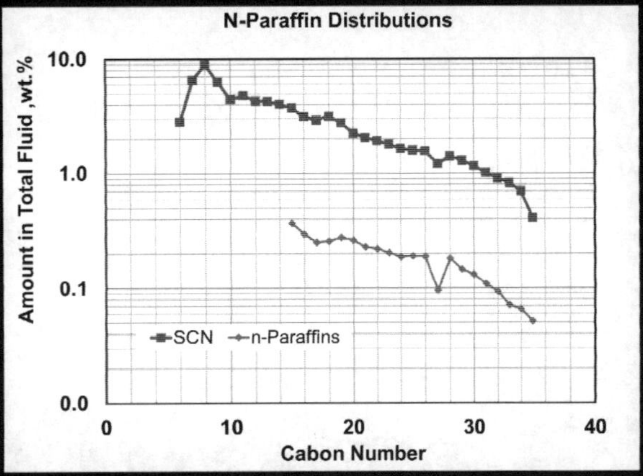

Input For Deposition Modeling

Assured Flow Solutions LLC

- Based on the same thermodynamics
  - Wax precipitation curve
  - Fluid viscosity

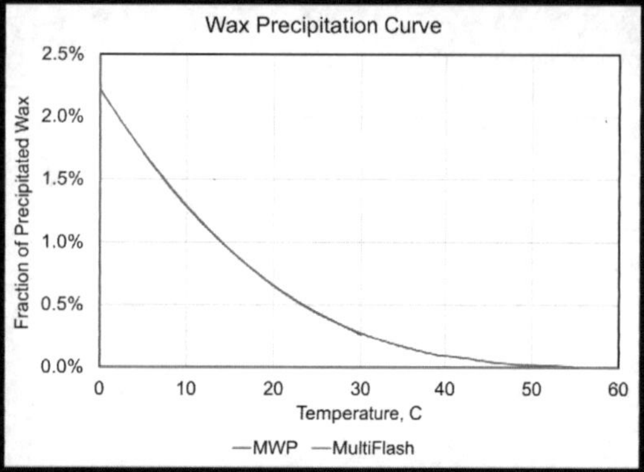

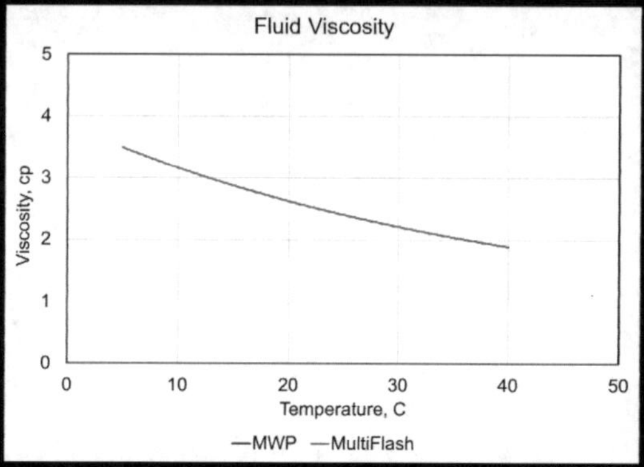

Results - Effect Of Oil Flow Rate

Assured Flow Solutions LLC

Experiment

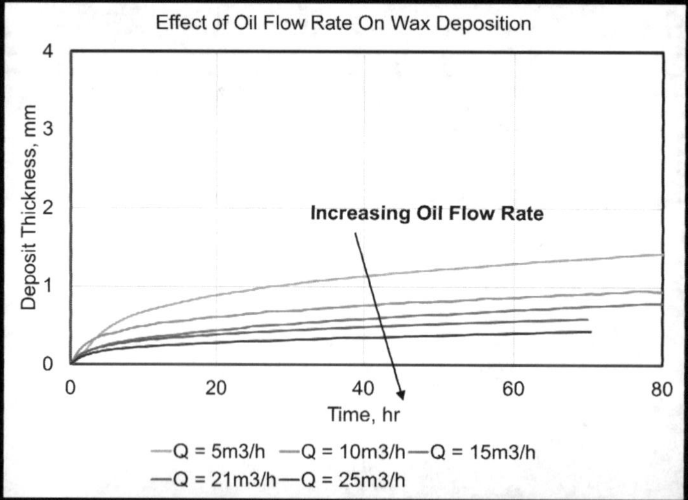

Deposit thickness decreases with increasing flow rate

Results - Effect Of Oil Flow Rate

Model Prediction (OLGA Wax)

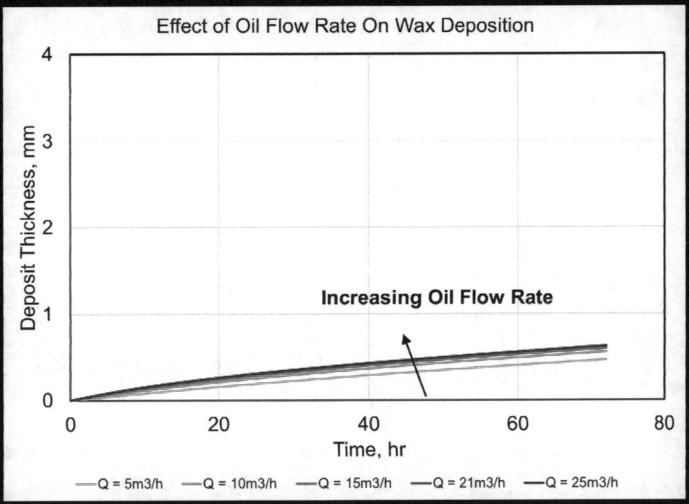

- Under-predicted in terms of Magnitude
- Trend opposite to experiments

Results - Effect Of Oil Flow Rate

Model Prediction (MWP)

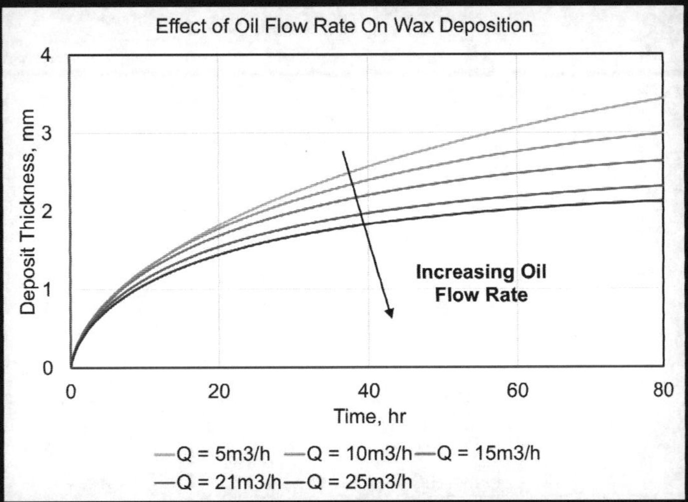

- Slightly over-predict in terms of magnitude
- Same trend as experiments

Results - Effect Of Coolant Temperature

Assured Flow Solutions LLC

Experiment

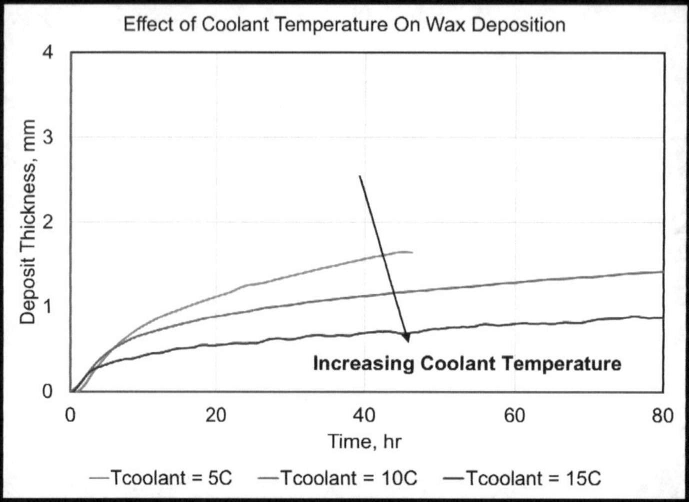

Deposit thickness decreases with increasing coolant temperature

Results - Effect Of Coolant Temperature

Assured Flow Solutions LLC

Model Prediction (OLGA Wax)

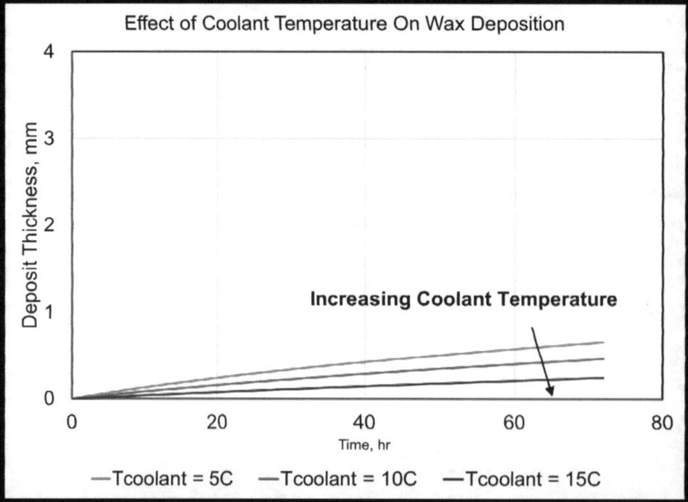

- Under-predicted in terms of Magnitude
- Trend same as experiments

Results - Effect Of Coolant Temperature

Model Prediction (OLGA Wax)

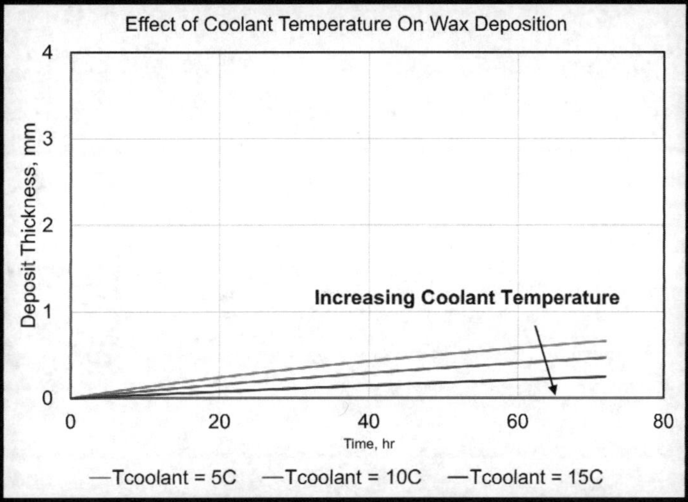

- Under-predicted in terms of Magnitude
- Trend same as experiments

Results - Effect Of Coolant Temperature

Model Prediction (MWP)

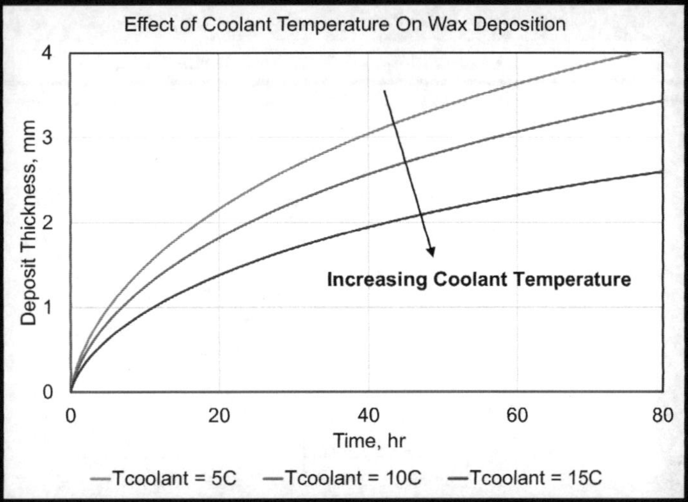

- Slightly over-predicted in terms of Magnitude
- Trend same as experiments

Summary

Assured Flow Solutions LLC

Operating Conditions	Experimental Trend	OLGA Wax	Michigan Wax Predictor
Increasing Flow Rate	Decreased Thickness	Increased Thickness	Decreased Thickness
Increased Coolant Temperature	Decreased Thickness	Decreased Thickness	Decreased Thickness

- Investigate the reasons for the difference:
  - Heat Transfer
    - OLGA Wax: 1D, MWP: 2D
  - Mass Transfer
    - MWP: Aging
    - OLGAWax: No Aging

Examination on Heat Transfer

Assured Flow Solutions LLC

- Pre-deposition Radial Temperature Profile

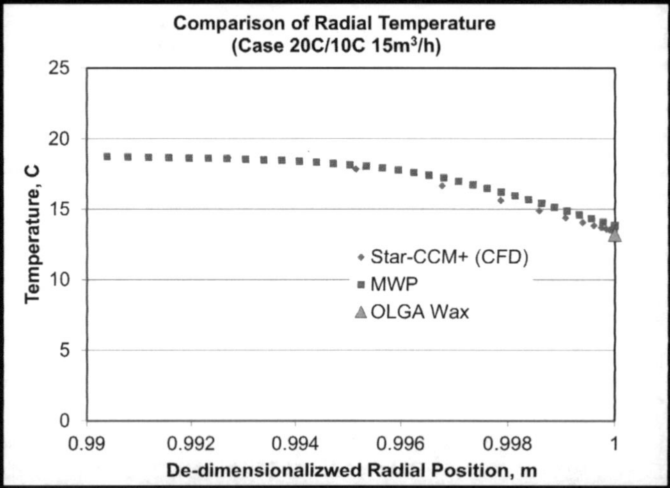

Temperature profile consistent with each other

Examination on Heat Transfer

Assured Flow Solutions LLC

➢ Pre-deposition Radial Heat Flux

Model	Star-CCM+ (CFD)	OLGA Wax	MWP
Heat Flux (W/m)	1009	928	944

Conclusion:

- Current industrial wax deposition models do not seem to accurately predicts the effect of operating conditions on wax deposition
- Heat transfer calculations seems consistent between three models
- The better performance in MWP is most likely due to its mass transfer calculations

Modeling Aging in Wax Deposition

Assured Flow Solutions LLC

➢ Aging: Increase of wax fraction in the deposit
- 4% wax in oil, 30%-50% wax in deposit
- Due to internal diffusion of wax in the deposit

➢ Why do we need to model aging?
- It affects wax mass balance (corrected in some of the industrial models)
- It affects diffusion of waxy components
- It affects the impact of insulation by the deposit layer

➢ How do we model aging?
- GC characterization of wax deposit
- Understand the solid-liquid equilibrium of waxy components in the deposit (*Sheng 2013, Energy and Fuels*)

Assured Flow Solutions LLC

Thank You!

352711 Development of Experimental and Modeling Methods to Understand Asphaltene Precipitation

Abstract:

Asphaltene Precipitation and subsequent deposition during oil production is a potential problem in many oilfields around the world, especially for light crude oils with relatively low asphaltene content, or in oilfields that are treated with miscible gas injection.

Different experimental procedures and modeling methods have been proposed by several research groups trying to understand the mechanisms of asphaltene precipitation and deposition. However, because of the complexity of these phenomena, there is still active discussion about the theories governing asphaltene stability.

In this work we present experimental evidence that supports the theory that precipitation and aggregation of asphaltenes is a multi-step process, where the former is driven primarily by thermodynamics whereas the latter is driven by kinetics. Under this multi-step mechanism, asphaltene precipitation, which leads to the formation of an impure asphaltene rich phase in liquid-liquid equilibrium with an asphaltene-lean phase, is a fully reversible process. On the other hand, from the precipitated phase, subsequent aggregation and aging leads to the formation of more solid-like structures. Furthermore, we will present experimental results that suggest that the currently available commercial technologies to detect asphaltene precipitation (i.e. NIR spectroscopy and High Pressure Microscopy) might not be appropriate to detect the exact point of asphaltene precipitation, but instead they give a combined reading of precipitation plus aggregation. For this reason, the results obtained using standard Solid-Detection-System are extremely sensitive to the depressurization rates.

These experiments are in conceptual agreement with the modeling approach for asphaltene stability based on the PC-SAFT equation of State. It was already proven that the PC-SAFT EOS can provide very accurate predictions of the phase stability of asphaltenes, bubble point of the oil, density of gas and liquid phases, GOR, composition of liberated gas and even asphaltene compositional grading that can induce in extreme cases the formation of tar-mats. In this work, we will present the progress of an enhanced characterization method that has been applied to various wells from three different fields in the Middle East. We found that it is possible to reduce from three to one the number of adjustable parameters for asphaltenes and we are still able to get comparable results for the phase stability of asphaltenes as well as the other properties mentioned above. The new method simplifies the fitting process and reduces its computational time significantly, without sacrificing the accuracy of the predictions.

With these results we aim to contribute to the understanding of the complex phenomena associated to the asphaltene phase behavior and the prediction of the occurrence and the magnitude of asphaltene precipitation and deposition.

352789 A Summary of the Solids Founds at the Modular Caustic-Side Solvent Extraction Unit Process

Abstract:

Since March 2008, the Actinide Removal Process-Modular Caustic-Side Solvent Extraction Unit (ARP-MCU) facility has successfully processed more than three million gallons of supernate. The process has experienced a few outages arising from plugged coalescers. Recently, the facility began an outage to perform maintenance activities aimed at extending the original life cycle for the facility in light of the future deployment of the Next Generation Solvent (NGS).

The outage included cleaning and maintenance of the centrifugal contactors. Inspection of the extraction centrifuges revealed the presence of accumulated solids in the bottom channels of Extraction Contactor SEP-401. SEP-401, the first centrifuge the supernate enters the solvent extraction process, is where radioactive supernate from ARP meets solvent scrubbed with 0.05 M nitric acid. The aqueous carry over from the solvent mixes with the supernate possibly resulting in a lowering of its pH to the point of solid precipitation.

Any large variance in the concentration of the aluminum and silicon concentrations, elements known to readily precipitate throughout the tank farm, may indicate precipitation driven by a mild pH swing. A similar deposit discovery (sodium aluminosilicate) was also found on September 2010 at the Salt Solution Feed Tank.

This poster presents a summary of the solids found throughout the different components that are part of the MCU at the Savannah River Site (SRS)

352853 Fighting Scales in Oil and Gas Production Systems – a Challenging Flow Assurance Case Study

Abstract:

Maturing oilfields has not just caused problems of large water handling but also demands novel flow assurance solutions which are practical and cost-effective. One of the major problems experienced in the oil production industry is the formation of mineral scale. Scales can form from the formation water as it undergoes changes in pressure and temperature, and tend to deposit on surfaces. Its formation causes flow reduction or even blocking of pipes, valves and other process equipment.

The main implications of the problem are:

- Economic; due to downtime equipment replacement, reduced or lost production
- Safety; Equipment failures due to under-deposit corrosion can cost lives.

Summarized, the impact of scales in Oil & Gas surface equipment can be serious and challenges in removal, prediction and prevention are numerous.

An approach was developed for utilizing prediction-based knowledge on the scaling problem to formulate a workflow for flow assurance monitoring. The workflow is also intended to aid in arriving at practical and cost-effective mitigation strategy for problems related to scaling in surface equipment. The following are the brief steps involved in the workflow:

- Identify & define the key Issues (such as production shortfalls, pressure anomalies, etc.)
- Define objectives and specify boundaries
- Model process (multiphase network model)
- Analyze Results and employ a variety of techniques to pin down the causes
- Conduct a structured assessment to determine/evaluate solutions
- Develop an action plan / process redesign for implementing the solutions

The root causes were confirmed as calcite scale deposition in flow lines. A detailed assessment was also performed to ascertain thermodynamic envelopes for solids deposition and identify specific hot spots related to pressure, temperature in the network to ensure trouble-free production.

The workflow suggested was valuable for troubleshooting bottlenecks of existing facilities & addressing production shortfalls. The lessons learned in this endeavor were also used to make better decisions for designing new projects.

352615 Computational Study of Oil-Water Wetting Regimes Established in Horizontal Flows: Wettability Maps and Corrosion

Abstract:

Wettability effects in pipelines carrying oil-water phases at varying proportions are crucial in undermining the net effects of corrosion in the system. Prediction of phase distribution, especially the information whether oil or water wets the pipe wall is an important factor, which can affect the corrosion mitigation strategy and increase confidence in the integrity of the pipeline. There are large knowledge gaps in this area of research and consequently only empirical criteria are used in the industry.

In the present work, detailed computational study involving, a turbulent oil-water multiphase flow system in a horizontal carbon steel pipeline, is carried out to predict different water wetting regimes as well as corrosion kinetics of emanating from the fluid-metal interaction. Computational Fluid Dynamics simulations include Eulerian multiphase models with appropriate sub-models for interfacial forces, turbulent interaction and population balance equations. Population balance techniques identify and follow the dispersed (water) phase dynamics such as droplet breakage and agglomeration in a detailed fashion. In addition, concentration equation for ferrous ion (Fe2+) is modeled to capture corrosion kinetics. The current study focuses on flow pattern computations for water cut ranging from 0 - 20%.

Our simulations captured various water wetting regimes inclusive of intermittent, fully wetting, dispersed and semi-dispersed water-in-oil behavior. Results were in good agreement with the experimental observation. Simulations clearly predicted that for given superficial water velocity, elevating the superficial oil velocity leads to transition of regimes from stable water wetting to intermittent water wetting and subsequently stable oil wetting scenario. Phasic fraction and velocity distribution at different flow cross-section reveal the flow modulations as a function of the turbulence inherent to the two-phase system. Corrosion computations clearly indicate a substantial generation of Fe2+ ion under stable water wetting regime while the concentration and corrosion levels deteriorate with intermittent water wetting, an effect well characterized in flow loop tests.

352338 Modeling of Conductive Antifouling Paints for Marine Applications

Abstract:

Protection from fouling organisms (biofouling) is a key function of coatings on marine hydrokinetic (MHK) devices. Accumulation of biomass on devices risks performance degradation and possible failure. At the present there are no proven reliable options for prevention of fouling in the marine environment with applications to match device lifetimes expected to be on the order of 10-20 years. Removal, cleaning, and re-application of short-lived antifouling paints is an expensive prospect. Long term solutions for antifouling protection need to be identified and tested for reliability. Electrified conductive paints have been shown to be effective in the prevention of marine biofouling and may be a viable solution for long term antifouling performance[1] . The present work seeks to further the understanding of the mechanism of action and to develop the engineering knowledge needed to enable application.

The mechanism of action has been proposed variously to be direct electron transport[1], electromagnetic field repulsion[2], and electrochemical reaction[3, 4]. Evidence suggests that electrochemical oxidation of chloride and bromide present in seawater produces strongly oxidizing agents which sanitize the surface being protected. These oxidizing agents are short lived in the environment with end states being the original chloride and bromide. Comparison of the experimentally determined minimum potential needed for biofouling protection with Tafel plots in sterilized and newly acquired seawater support this hypothesis, as does cyclic voltammetry.

Chloride oxidation at graphitic anodes has been studied in highly saline water at high over-potential for the chloralkali process. The salinity, current density, product concentration, and pH conditions of the chloralkali process are vastly different from the conditions at a protected surface in the natural marine environment with current density on the order of $20\mu A/cm^2$. Determination of reaction products and reaction selectivity under the applicable conditions is necessary to allow design of electrochemical antifouling systems. The present work utilizes a well characterized millimeter scale laminar flow electrochemical reactor and amperometric methods to determine reaction products, which yields reaction selectivity.

In order to evaluate the longevity of conductive antifouling coatings in marine applications accelerated aging methods are required. Standard procedures for rapid aging of marine paints involve exposure to ultraviolet light, moisture, temperature swings, and salt fog[5]. The two primary concerns for MHK applications are adhesion and electrochemical performance. The existing test regimens do not address electrochemical performance; therefore in addition to the standard tests accelerated aging methods using high over-potentials and current density have been developed. The electrode selectivity is characterized throughout the aging process, which enables prediction of performance losses.

Aged coatings have also been deployed in the marine environment to validate performance after aging. A standard procedure for washing test coupons to remove loosely adhered debris using consistent shear stress followed by visual and microscopic inspection has been developed as a means to standardize determination of antifouling efficacy.

[1] T. Matsunaga, T. Nakayama, H. Wake, M. Takahashi, M. Okochi, and N. Nakamura, "Prevention of marine biofouling using a conductive paint electrode," *Biotechnol. Bioeng.,* vol. 59, pp. 374-378, // 1998.

[2] T. Nakayama, H. Wake, K. Ozawa, N. Nakamura, and T. Matsunaga, "Electrochemical prevention of marine biofouling on a novel titanium-nitride-coated plate formed by radio-frequency arc spraying," *Appl. Microbiol. Biotechnol.,* vol. 50, pp. 502-508, // 1998.

[3] R. E. Perez-Roa, M. A. Anderson, D. Rittschof, C. G. Hunt, and D. R. Noguera, "Involvement of reactive oxygen species in the electrochemical inhibition of barnacle (Amphibalanus amphitrite) settlement," *Biofouling,* vol. 25, pp. 563-71, // 2009.

[4] J.-R. Huang, W.-T. Lin, R. Huang, C.-Y. Lin, and J.-K. Wu, "Marine biofouling inhibition by polyurethane conductive coatings used for fishing net," *J. Coat. Technol. Res.,* vol. 7, pp. 111-117, // 2010.

[5] *Paint and Coating Testing Manual : 15th edition of the Gardner-Sward handbook*: ASTM International, 2012.

2014 AICHE Spring Meeting
March 30 to Apr 03, 2014

Burns & McDonnell
1700 West Loop South, Suite 300
Houston, TX 77027

Authors/Presenters:
Mark W. Lockhart, P.E., Process Technology Manager, Chemicals and Petrochemicals
Burns & McDonnell, Houston, TX
832-214-1975

Elliott Robertson, P.E., Senior Process Engineer, Chemicals and Petrochemicals
Burns & McDonnell, Houston, TX
832-214-2916

Expanded Abstract:

"Shale Gas Boom – A New Era of Technology for Ethylene and Ethylene Derivatives"

There is a new era in the U.S. with the rise of the Shale Gas Boom that has emerged in this country. Large production increases are occurring from shale gas plays ranging from C1's – C4's, C5's/condensate and "tight oil". The plentiful supply of C1's – C4's versus current demand is driving prices for these molecules down. The recent result is directionally lower prices for methane, ethane, propane and butane as well as heavier products.

The plentiful supply and low prices for these building block molecules is having a marked effect on downstream industries that use the molecules either as fuel for energy or for raw materials as feeds. Steam crackers are moving to maximally change feedstock from naphtha's and gas oils to ethane to make ethylene and recently experiencing very attractive margins as a result.

Ethylene yield is very high when using ethane as feed which results in very favorable economics given the low price of ethane that has been prevalent over recent years. Ethylene producers have already made moves to feed as much ethane as possible within the limits of their existing production facilities. Ethylene producers are also moving to debottleneck existing plants including various debottlenecks that have already been announced. Additionally, as many as seven (7) new grass-roots facilities have been announced or are under consideration primarily along the gulf coast. Although, at many as two new ethylene plant projects are under review to be located in the North East to enable taking advantage of ethane production from the Marcellus Shale Play located in the states of Ohio, Pennsylvania and West Virginia.

Ethylene capacity totaling over 7 mm MTons/yr have been announced or are under study between debottlenecking of existing plants and new grass-roots ethane crackers. The debottlenecking projects can be accomplished in a shorter timeframe with the capacity increases continuing through 2013 to 2015. The new grass-roots capacities are slated to come on-line through 2016 and 2017. Several of the new ethane crackers will also install additional polyethylene capacity as part of the new facilities.

Historically, naphtha and gas oils have been used as feed to steam crackers. These feeds typically come from processing crude oils in refineries and, as such, the pricing for naphtha and consequently ethylene is directly related to the price of crude. Ethane production from shale gas plays provides crackers a new source of feed that is not related to the refining of crude oil or world-wide oil markets. This represents a significant paradigm shift in the traditional economics for ethylene production.

Ethylene is a major building block for numerous chemicals and petrochemicals. Consequently, the shale gas renaissance is affecting a broad spectrum of the chemical industry.

Shifting ethylene production from naphtha and gas oil to ethane as feed has a dramatic effect on the production of other byproducts that traditionally come from steam crackers. The yields of propylene, C4's, C5's and heavier products that are produced when feeding ethane are very low.

The reduction in propylene production is affecting the economics of producing propylene through the "on purpose" route. Numerous propane dehydrogenation projects are already in design with others being announced. This is just one example of the changing markets and the necessary transition that is taking place in the U.S. as the markets are in transition.

The total spending for the chemical/petrochemical industry for the period 2013 to 2030 is now estimated at approximately $120 billion. There is an anticipated sharp increase in spending starting in 2014 to peak in 2016/2017 at over $13 billion. The total spending for the period 2013 to 2017 totals nearly $48 billion.

With the advent of the announced ethylene plants slated to come on line by 2017, ethylene will be in greater supply. The economics of supply and demand suggest an oversupply of ethylene produced in the U.S. which will result in lower ethylene prices. A lower ethylene price has the effect of swinging economics in favor of producing more derivatives made from ethylene which will have to be exported from the U.S. to the global market place in order to maintain a supply and demand balance.

It is estimated that along with ethylene production various ethylene derivatives including polyethylene, ethylene oxide, ethylene glycol, EDC, VCM, PVC, among others will also increase. This has the effect of increasing the capital spending through 2020 - and beyond.

Where will all this production go? It is forecasted that the U.S. will be increasing net exports of the ethylene derivatives.

Lower energy cost along with cheap ethane has caused a marked shift in where U.S. stands on the global ethylene cost of production curves. An impressive result of the Shale Gas Boom and consequent lower cost of energy and cheap ethane has been the effect on U.S. cost of production for ethylene on the global cost curve. Comparing the 2007 and 2012 cost of production curves it is clear that the shift has now brought the U.S. much closer to the lowest cost producers including Canada and the Middle East.

Ethylene Dichloride (EDC) is one of the derivatives of ethylene that is forecasted to experience capacity increases in the U.S. EDC is used to produce Vinyl Chloride Monomer (VCM) which in turn is used to produce PVC. The U.S. is likely to increase export of PVC for use in home building and plumbing products world-wide. Burns & McDonnell has chosen to highlight the EDC -> VCM value chain in this presentation.

The process to produce chlorine is very energy intensive at 20,200 Btu/lb. Reduced U.S. energy costs represent a significant advantage in lowering the cost of production for chlorine which is another advantage in the EDC, VCM and PVC value chain. There are two widely accepted methods for producing EDC: direct chlorination and oxychlorination.

In direct chlorination, chlorine and ethylene are reacted to form the EDC molecule such that two chlorine molecules are used for each molecule of ethylene. The typical reaction is between the dissolved gases in liquid over an iron chloride catalyst at 15 to 75 psig pressure and 200 to 300 deg F. The use of high purity reactants is typical to avoid the formation of a broad range of impurities which may further complicate EDC purification. An excess of chlorine is typical to ensure complete ethylene conversion.

In oxychlorination, the reaction occurs over a copper chloride impregnated alumina catalyst in the liquid phase at a temperature of approximately 300 deg F and a pressure of up to 75 psig. The typical fluid bed technique offers heat transfer, prevents the occurrence of hot spots and allows more efficient catalyst regeneration. Ethylene conversion of 93 to 97% can be achieved with selectivity in EDC of 91 to 96%.

The EDC purification section removes trace amounts of impurities in the furnace feed which can inhibit the cracking reactions and cause fouling. In addition, EDC must be dry to prevent excessive corrosion in equipment and piping downstream of the reactor. Therefore the combined EDC from three different sources must be purified to greater than 99% EDC from direct chlorination, EDC from oxychlorination and EDC recovered and recycled from the cracking step. This combined EDC is washed with water and scrubbed with caustic to remove chlorine, HCl and other water soluble impurities including chloral. The crude EDC and unconverted EDC from the VCM column are combined and dried and purified in two distillation columns. In the first "Lights Column", water and lower boiling point impurities are separated as the overhead product. The bottoms product is fed to a "Heavies Column". Pure EDC of approximately 99% purity is taken as the overhead product while the heavy impurities leave the column as the bottoms product.

The pure EDC is forwarded to a furnace for thermal cracking of EDC to vinyl chloride. A typical furnace consists of a radiant section, a convection section, and a stack. The furnace fire box operates at 950 deg F. The convection utilizes combustion gases to preheat the feed before entering the firebox.

The vinyl purification section consists of two distillation columns in series which are used to separate vinyl chloride from EDC, HCl and remaining by-products. The first column ("HCL Column") distills the HCL mixture to a pure overhead product which is recycled to the oxychlorination reactor. The HCL Column is a trayed column and operates at a pressure of 135 psig. The bottoms product of the HCL Column is fed to the second column ("VCM Column") which operates at 65 psig. A VCM product of greater than 99 wt% is produced as the overhead product of the VCM Column. The bottom of the VCM Column is recycled to the Lights Column for re-purification.

A brief overview of the economic analysis for EDC/VCM plant investment indicates that the total installed cost (TIC) for a 300,000 metric ton/yr facility is in the range of $450MM to $500MM. With current EDC pricing of $245-275/ton and assuming an ethylene and chlorine cost of $1320-1362/ton and $275-310/ton, respectively and assuming the energy, labor, overhead and tax contributions are in the range of $70-85/ton results in a payback of less than five years. These rough economics do not include cash costs for land, start-up nor the capital investment required for utilities and infrastructure.

Constructing a new EDC plant at an existing facility where existing infrastructure and utilities are present will make the investment more attractive than a pure green field site. Debottlenecking an existing EDC facility is likely yield an ROR that represents another attractive investment.

Furthermore, installing or debottlenecking existing VCM and PVC facilities make this value chain one of keen interest in the coming low cost ethylene environment that is predicted.

In conclusion:

- The Shale Gas Boom is resulting in high production for C1's – C4's, which has resulted in reduced prices for these chemical and petrochemical building blocks.
- Several new grass-roots ethane crackers and ethylene plant debottlenecks to increase ethane feed capacity have been announced.
- Increased ethylene production in the U.S. is forecasted to lower ethylene prices in the long run.
- Ethylene is a central building block in producing various chemicals and petrochemicals.
- Therefore, a large sector of the U.S. chemical industry will be affected.
- Ethylene derivatives benefit from reduced ethylene prices.
- U.S. exports of ethylene derivatives are forecasted to increase to supply the world-wide market.
- The U.S. can compete well in the world chemical and petrochemical market place with the advent of the Shale Gas Plays.
- Chlorine cost of production is and will continue to benefit from the low energy prices in the U.S.
- EDC and VCM are examples chemicals that are forecasted to benefit as more ethylene production comes on line along with consequent lower ethylene pricing.
- The economics of producing EDC and VCM in a low cost ethylene environment is compelling.
- The EDC->VCM->PVC value chain will benefit. Export of these chemical products to the world-wide market is forecasted to increase in the future.

The Shale Gas Boom has shifted the U.S. energy picture as well as the outlook for chemical and petrochemical production. The U.S. is once again a key actor and leader in the world chemicals stage.

This truly is in a new era of Technology!

347676 Opportunities in Chemical Engineering Being Created by New US Shale Gas Supplies

Abstract:

Recent technical breakthroughs in extraction techniques have allowed the combination of hydraulic fracturing with long horizontal wells to enable the release of unconventional hydrocarbon sources. The ability to tap into these previously unreachable shale gas resources has completely transformed the US natural gas and crude oil supply and demand outlook. These recent technology breakthroughs have resulted in natural gas production in the United States increasing by 25% over the last 5 years, largely due to shale gas production. New US natural gas supplies are playing a key role in "turbocharging" US manufacturing over the next 10 to 20 years, creating an industrial renaissance. This paper summarizes potential opportunities in chemical engineering that could arise as a result of new shale gas production.

350754 Shale Gas Boom and Tight Oil - a New Era for the Refining Industry

Tuesday, April 1, 2014: 3:30 PM
Oak Alley (Hilton New Orleans Riverside)

Mark W. Lockhart, Refining & Chemicals, Burns & McDonnell, Houston, TX and **Dominic Varraveto**, Process Engineering, Burns and McDonnell, Kansas City, MO

Production of Tight Oil from the Shale Gas Plays in the U.S. is having a profound effect on the Refining Industry. Refiners are maximizing the amount of advantaged, "tight oil", they can feed their crude units some with geographic locations that allow them to capitalize the most on the sometimes attractive spreads provided by what is mostly light and sweet crude. New topping columns, crude units and condensate splitters are being constructed in various refineries in the U.S. This paper provides an overview of some of the studies and projects that are taking place and provides an overview across the spectrum of possibilities that refiners are facing. Burns & McDonnell has much experience with actual projects that will be highlighted to provide an exciting overview of this activity.

347649 Economically Treating Sour Gas from Tight Oil Formations

Abstract:

GT-SPOC – Economically treating sour-gas in the shale plays

In the shale plays, there are pockets of sour gas, or sour associated gas. Treating this gas to pipeline specifications with amine units gets the gas to pipeline specifications. However, operators have to deal with the resulting acid gas stream that cannot generally be vented due to permit limits on sulfur emissions. If the gas is not very sour, sulfur scavengers are the obvious solution, however, as the gas contains more than 0.1 ton/day of sulfur, the treating method of choice is not as clear.

GT-SPOC offers an alternative to the conventional furnace based "modified Claus" method for sulfur recovery. The technology is very suitable in the 0.5 – 20 ton/day sulfur production range, which is typical of the sulfur rates we see in the shale plays.

The approach improves upon the Claus process which is used to produce >95% of sulfur around the world, by using a patented catalyst at the front of the plant for SO2 production. The catalyst utilizes constituents that address typical operator issues relating to heavy hydrocarbon contamination of the gas, start-up and shut-down issues, sour-gas composition variations and turndown. The presentation will discuss the capital and operating economics, as well as specific technical aspects of the technology that allow modularization, reduced footprint, and a tremendous reduction in capital cost.

352037 Models Describing the Thermal Recovery as a Productive Option for the Heavy Oil Reservoirs

Abstract:

The crude oil requirement is increasing relatively at a higher rate in comparison to the amount of new crude reserves being identified. To meet the requirement, the recovery from the available crude reserves should be maximized. After the primary recovery, there is a still a good proportion of crude is still left in the reserve. There is a need for the secondary and tertiary methods for the recovery of the left out crude.

Thermal methods are one of the successful techniques which can be implemented to enhance the recovery. I have been tested since 1950's, and they are the most advanced among all the enhanced oil recovery methods available, as far as field experience and technology are concerned. The best utilization of thermal methods can be in the reserve left with heavy oils (of 10-20° API) and tar sands (≤10° API). Thermal methods provide a source of energy in the reservoir resulting in the large reduction in the viscosity of the heavy oil left. Decrease in the viscosity will result in a higher mobility ratio.

Thermal methods have been highly successful in Canada, USA, Venezuela, Indonesia and other countries. The general classification of Thermal enhanced oil recovery can be broadly classified in terms of hot water, steam, in-situ gasification or electrical heating. Among these methods the steam based methods are more successful commercially than others. In the present study, we will be discussing about the various models used to understand the behaviours of the reservoir in response to the thermal enhanced oil recovery methods. We will be discussing about what are the type of simulators being used to identify the behaviour.

Fire flooding as a budding source of enhanced oil recovery

- by KHUSHBOO SINGH (f2011409@pilani.bits-pilani.ac.in)

Fire flooding is a thermal method of oil recovery in which compressed air is injected into an injection well and the petroleum in the reservoir surrounding the injection well is ignited to establish a combustion zone or fire front .This fire front propagates in a radially expanding manner away from the injection well. The heat reduces the viscosity of oil, cracks some higher molecular weight hydrocarbons into smaller ones, vaporizes some lighter hydrocarbons to help miscibly displace oil and also produces steam that may steam distill the trapped oil .Thus, along with decreasing the mobility ratio, it also produces improved oil. Also, the gases produced due to combustion and nitrogen (in air) displace the oil towards reservoir away from fire front again enhancing production .Other methods of EOR cannot be used for highly viscous oils whereas thermal recovery or say, fire flooding can be used for both high and less viscous oils. Fire flooding also offers many advantages over the other method of thermal oil recovery i.e. steam flooding. Lot of fresh water and capital is used to produce steam for steam flooding whereas in this there is just the need to push the compressed air and ignite the oil thus, lowering the cost of production .Also, it has the potential to operate in wide variety of reservoirs that other methods. For example, lower in pressure (as it lowers the viscosity (improved mobility)as well as produces gases that supply pressure), containing more shale , lower in quality and thinner (as it requires less volume of air for producing the same amount of oil than other methods which employ gases like carbon dioxide method of oil recovery)and deeper (heat losses occur in steam flooding).But it also has some limitations like it can be applied only to reservoirs of high permeability , high porosity and oil saturation and has poor sweep efficiency in thick formations .Also, if excessive coke is deposited the rate of advancement of combustion zone will be slow and more air will be required. It also produces flue gases which cause environmental problems. Along with this it also faces operational problems like corrosion due to low pH hot water, serious oil water emulsions, and increased sand productions, deposition of carbon or wax and pipe failures to high temperatures.

Now however, oxygen enriched air or pure oxygen is being used as the combustion supporting medium for which about one-fifth as much volume is required for the injection process as compared to compressed air since nitrogen which accounts for about 80% of the volume of the compressed air is eliminated. Thus, the use of the pure oxygen in the process instead of air not only drastically reduces the volume of the combustion supporting medium required but also provides a decrease in the fluid volume following combustion which generally increases the injectivity of the reservoir matrix in the combustion zone .Also, it produces carbon dioxide which dissolves in oil and reduces its viscosity. Thus, one of the major limitations of fire flooding i.e. it requirement of high permeability in reservoirs is eliminated in this way .After an initial level of in situ combustion the production well is partially choked or shut in until the bottom hole pressure increases to a substantial fraction of the injection pressure .The production well can then be opened to a lower back pressure level which results in an immediate acceleration of fluid flow under the increases oil pressure gradient and leads to enhanced recovery. Thus, pressurization and producing cycles can be repeated using intervals found to be most effective for the particular system. Also, water or steam or mixtures of steam and carbon dioxide (or low molecular weight C3-C8 hydrocarbons) can be injected intermittently or subsequent to injection of combustion supporting medium to enhance production. Further, relatively heavy viscous crude oil from a subterranean stratum can be produced by the use of in situ combustion. In conventional direct drive or forward underground in-situ combustion, when applied to certain oil-bearing strata, particularly those containing a heavy viscous crude oil, it becomes impossible to continue to support an expanding combustion zone because of the formation of a viscous liquid bank of oil in advance of the combustion zone. This viscous liquid bank, substantially completely plugs the stratum and prevents continued injection of air to feed the combustion zone .So, now the oil is ignited in the stratum to be produced around a single well penetrating same and air is injected through the well to expand and move the ignited area (combustion zone) into the stratum away from the well whereby a liquid bank of viscous oil builds up in the stratum beyond the combustion zone so as to substantially plug the stratum and build up substantial pressure above the stratum pressure at the time of initiation of the process, and thereafter terminating the injection of air and opening the well to production so that the fluid pressure in the stratum beyond the combustion zone forces liquid oil from the liquid bank back into the combustion zone and the hot burned-over area behind the combustion zone so as to force the upgraded oil into the well from which it is recovered by conventional methods. After a substantial period of production during which the pressure-build up in the stratum during the injection period forces oil in upgraded condition into the well, the stratum is again ignited adjacent the well to establish a combustion zone which feeds on the oil filling the pores of the stratum at the end of the first production phase. In the event the stratum is below combustion supporting temperature, the ignition is effected in conventional manner as by heating with an electric igniter, an ice charcoal pack, a gas fired heater, or similar device and injecting air into the hot stratum. Also, recovery can be improved using water containing surfactant and alternate slugs of air during forward combustion. In an underground in situ combustion process for the recovery of oil from formations having high permeabilities, after combustion has been initiated, water containing a surfactant and air can be injected under conditions such that the water upon contact with the hot formation vaporizes, creating slugs of condensable foam in an amount sufficient to generate a significant resistance factor of at least 5, which improves volumetric sweep efficiency by retarding the mobility of the steam and subsequent slugs of air. Air and water, including a surfactant, can be injected simultaneously as a mixture or as alternate slugs of water containing a surfactant and air. It also provide a method for temporarily or partially plugging the more permeable strata and a process for producing a well by selectively blocking and plugging highly permeable sections of subterranean oil-bearing formations with a self-destructive foam generated in situ exhibiting a significant resistance factor. Thus, as we saw in situ combustion can be applied to a wide array of reservoirs whether it be thin, low pressure or deep reservoirs. In fact it is the only thermal method that can presently be applied to deep reservoirs, though deep down steam generation is being tested. It can be used at any stage of reservoir depletion. It can be used in special situations such as offshore or in arctic regions. Because of the lack of heat losses at the surface and in the injection wells, it is the most thermally efficient thermal recovery method. The inject ant (air) is readily available. Combustion allows wider well spacing than steam. Economic results are comparable to those of steam injection.

Several aspects of operating ISC projects are important. First is the large compression ratios and associated costs required to inject air into the formation. Second is planning and design requirements for a combustion project that are more difficult than steam. Third is extensive laboratory work to assess fuel availability, air requirements and burning characteristics of the crude that are required before designing in situ combustion projects. Fourth is the high degree of technical sophistication and careful monitoring needed to ensure proper operation of project Fifth is the limitation of numerical simulation and other techniques that make predictions of recovery more difficult than most other enhanced oil recovery methods .As mentioned before, considerable improvements in the application of in situ combustion have been made since the early projects .New developments like using oxygen enriched air, injecting water containing surfactant and alternate slugs of air during forward combustion etc improve the efficiency of fire flooding as well as eliminate the problems faced during its application.

352165 Design of Surfactant-Only Chemical Flooding in a Low Perm, High Tds Reservoir: Single Well Test

Abstract:

Low permeable oil reservoirs are often characterized by slow waterflood injectivity and low fluid productivity. For low perm formations, the permeability appears to be the limiting factor to deploy Surfactant-Polymer (SP) and Alkaline Surfactant Polymer (ASP) flooding for improved oil recovery. Injection of carbon dioxide has shown promising results in low perm formations and is commonly used in these fields when sufficient CO_2 can be obtained. However, even when sufficient CO_2 can be sourced, initial capital investments for CO_2 injection may be quite high and may limit its application for some mature fields. In this paper we report a single-well test of an improved surfactant-only formulation in a low permeability, high TDS reservoir. The target single well, SEH1, is located in south central of Oklahoma. The perforated zone is located 6,000 feet below the surface with an on-going waterflood. The average permeability of the formation is 14 mD, ranging from 0.2 to 44 mD. The produced brine contains total dissolved solids (TDS) at 102,300 mg/L and the reservoir temperature is 50 °C (123 °F). The reservoir crude exhibits a viscosity of 23 cp at reservoir conditions, presenting further challenges for an EOR process without mobility control using polymer. For the test, the hydrophilic-lipophilic difference (HLD) equation was used to predict possible salt-tolerant surfactant formulations, based on site-specific fluid properties and reservoir conditions. Among the binary and ternary surfactant mixtures developed, the surfactant formulations were further optimized by introducing low concentrations (between 400 to 600 mg/L) of hydrophilic linkers to achieve a stable surfactant solution in the reservoir brine and ultra-low interfacial tensions (< 0.008 mN/m) between site crude and produced brine. In laboratory testing injection of three pore volumes of a surfactant-only formulation in sand packs and core flood devices achieved tertiary oil recovery ranging from 45% to 70% of the residual oil (Sor) after water flooding. The levels of surfactant loading were varied from 0.23

wt% to 0.92 wt% to assess the impact of surfactant concentration along with slug size. The field single well test used a ternary surfactant formulation (i.e., a mixture of alkoxylated alcohol sulfates/ethoxylated alcohol sulfates/blends of monoalkyl and dialkyl diphenyloxide disulfonates) and was conducted in April-May of 2013 to further confirm laboratory results at the SEH1 site. The 3-week single-well test was a technical success. A surfactant only formulation was injected at 0.69% followed by injection of reservoir brine without mobility control. Based on the pre- and post-chemical tracer test results (i.e., use of ethyl formate as partitioning tracer plus other conservative tracers), approximately 90% of the residual oil at the target zone was mobilized. This field experience will be built into a design protocol using single-well tests to mitigate the risks of pilot-scale chemical flooding targeted at mature fields in the U.S. mid-continent region.

352753 Foam Mobility Control: An Enhanced Oil Recovery Technique

Abstract:

In Enhanced Oil recovery techniques a significant fraction of the oil-in-place is left in the ground after primary or secondary recovery due to various reasons. Gas injection is one method used to extract these oils. There are basically four types of gases used in gas injection: natural hydrocarbon gas, inert gas, carbon dioxide, and steam; but in all cases there is the problem of gravity override.

Gravity override meaning the preferential movement of the injected gas through the upper part of the reservoir due to the lighter density of gas compared with oil. Gravity override causes early gas breakthrough in production wells and hence, reduced oil recovery. The steam is injected into the ground to reduce the viscosity of the oil and to make the oil more mobile. Unfortunately, due to gravity over ride, as steam rises to the top of the reservoir. Together with reservoir heterogeneity, gravity override results in early steam production and reduced volumetric sweep efficiency. Hence recovery is not improved to the degree desired. The injection of a permeability blocking agent would do the required. Various scientist have reviewed the selective blockage of fluids in thermal recovery projects and concluded that foam is best suited for this purpose. The foam is found to be promising to work in alkaline/surfactant processes to replace or supplement polymer as mobility control agent. The present study deals with the advancements in the foam mobility control for the recovery of unrecovered crude oil from the reserve.

AUTHOR INDEX

AUTHOR INDEX